T0205806

Stem Cell Surgery Trials in Heart Failure and Diabetes

Jeffrey N. Weiss
Author

Stem Cell Surgery Trials in Heart Failure and Diabetes

A Concise Guide

 Springer

Jeffrey N. Weiss
Parkland, FL, USA

ISBN 978-3-030-78009-8   ISBN 978-3-030-78010-4   (eBook)
https://doi.org/10.1007/978-3-030-78010-4

This Springer imprint is published by the registered company Springer Nature Switzerland AG
The registered company address is: Gewerbestrasse 11, 6330 Cham, Switzerland

Preface

Parts I–IV of this book identify and summarize the current worldwide stem cell trials for heart failure. Only studies using non-expanded stem cells and not being used in conjunction with a proprietary or non-proprietary treatment are included. All studies are Institutional Review Board (IRB) approved and registered with clinicaltrials.gov.

The studies are grouped by the type of stem cells utilized and divided between US and non-US studies. Only "recruiting"; "not yet recruiting"; "active, not recruiting"; and "enrolling by invitation" studies are included. Studies that were listed as "suspended," "terminated," "completed," "withdrawn," or "unknown status" were not included. Each submission was individually reviewed, and errors were corrected.

Parts V–XV identify and summarize the current worldwide stem cell trials in patients with diabetes mellitus. All studies are Institutional Review Board (IRB) approved and registered with clinicaltrials.gov.

The studies are grouped by the type 1 or type 2 diabetes diagnosis and the geographic location of the study. Only "recruiting"; "not yet recruiting"; "active, not recruiting"; and "enrolling by invitation" studies are included. Studies that were listed as "suspended," "terminated," "completed," "withdrawn," or "unknown status" were not included. Each submission was individually reviewed, only pertinent studies were included, and errors were corrected.

Having a listing and description of studies pertaining to the use of stem cells in the treatment of heart failure and in diabetes provides the physician with a handy knowledge of IRB-approved studies in these fields.

Parkland, FL, USA   Jeffrey N. Weiss

Contents

Part XVI Other Thoughts

Part I
Heart Failure (United States)

Chapter 1
Bone Marrow-Derived Mesenchymal Stem Cells in Improving Heart Function in Patients with Heart Failure Caused by Anthracyclines

Brief Summary:

This randomized pilot phase I trial studies the side effects and best method for the delivery of bone marrow-derived mesenchymal stem cells (MSCs) to improve heart function in patients with heart failure caused by anthracyclines (a type of chemotherapy drug used in cancer treatment). MSCs are a type of stem cell that can be removed from the bone marrow and grown into many different cell types that can be used to treat cancer and other diseases, such as heart failure. Bone marrow-derived MSCs may promote heart muscle cell repair and lead to reverse remodeling and ultimately improve heart function and decrease morbidity and mortality from progression to advanced heart failure.

Condition or disease	Intervention/treatment	Phase
Cardiomyopathy Heart failure	Biological: Mesenchymal stem cell transplantation Drug: Standard of care for heart failure	Phase 1

Primary Objectives:

I. To demonstrate the safety of allogeneic human mesenchymal stem cells (hMSCs) administered via intravenous infusion in patients with left ventricular (LV) dysfunction and heart failure secondary to chemotherapy with anthracyclines.

Secondary Objectives:

I. To demonstrate the efficacy of allogeneic hMSCs administered via intravenous infusion in patients with left ventricular dysfunction (left ventricular ejection fraction [LVEF] <40%) and heart failure secondary to treatment with anthracyclines.

MD Anderson Cancer Center, Houston, Texas, United States
ClinicalTrials.gov Identifier: NCT02408432

Outline: Patients are randomized to 1 of 2 treatment arms.

Arm I: Patients receive allogeneic hMSCs intravenously (IV) over 10–20 minutes once weekly for 4 weeks and standard-of-care treatment for heart failure.

Arm II: Patients receive only standard-of-care treatment for heart failure.

After the completion of study treatment, patients are followed up monthly for 6 months and then at 12 months.

Study Type:	Interventional (Clinical Trial)
Estimated Enrollment:	45 participants
Allocation:	Randomized
Intervention Model:	Parallel Assignment
Masking:	None (Open Label)
Primary Purpose:	Treatment
Official Title:	Intravenous Administration of Allogeneic Bone Marrow-Derived Multipotent Mesenchymal Stem Cells (MSCs) in Patients with Recent Onset Anthracycline-Associated Cardiomyopathy
Actual Study Start Date:	January 11, 2016
Estimated Primary Completion Date:	January 31, 2023
Estimated Study Completion Date:	January 31, 2023

Arm	Intervention/treatment
Experimental: Arm I (hMSCs) Patients receive allogeneic hMSCs IV over 10–20 minutes once weekly for 4 weeks and standard-of-care treatment for heart failure	Biological: Mesenchymal stem cell transplantation Undergo mesenchymal stem cell infusion Drug: Standard of care for heart failure Both groups will also receive standard-of-care treatment for heart failure. The study doctor will tell you which drugs you should take, their risks, and when you should take them Drug: Standard of care for heart failure Both groups will also receive standard-of-care treatment for heart failure. The study doctor will tell you which drugs you should take, their risks, and when you should take them
Active comparator: Arm II (standard-of-care treatment) Patients receive only standard-of-care treatment for heart failure	Drug: Standard of care for heart failure Both groups will also receive standard-of-care treatment for heart failure. The study doctor will tell you which drugs you should take, their risks, and when you should take them

Primary Outcome Measures:
1. Adverse events of allogeneic human mesenchymal stem cells (hMSCs) determined by CTCAE version 4 [time frame: up to 6 months]

The 90% criterion is achieved if the first 2 patients on investigational arm die and the single patient assigned to the control arm is still alive. It will also be achieved if 4 of the first 10 investigational patients die and there are no deaths among the 5 patients in the control arm. However, the probability of an increased death rate would be less than 90% if 1 of the 5 control patients had died. As a third example, the 90% criterion would be achieved if there would be 7 deaths among 20 investigational patients and 1 death among 10 control patients, but not if there would be 6 deaths among the 20 investigational patients. Statistical analyses of safety will be descriptive.

Secondary Outcome Measures:
1. Change in the left ventricular ejection fraction (LVEF) [time frame: 6 months]

Change in the left ventricular dysfunction (LVEF) from baseline to 6 months after randomization, expressed as a percentage. The comparison will be between the two groups of patients.

Ages Eligible for Study:	18 Years to 80 Years (Adult, Older Adult)
Sexes Eligible for Study:	All
Accepts Healthy Volunteers:	No

Inclusion Criteria:
- Patients with LVEF = <40% from treatment with anthracyclines for all malignancies at any dose at any time without evidence of other causes of cardiomyopathy
- Documented New York Heart Association (NYHA) classes I, II, and III
- Been treated with appropriate maximal medical therapy for heart failure
- Able to perform the 6-minute walk test
- Patient or legally authorized representative able to sign informed consent
- Patients with persistent LV dysfunction 90 days after discontinuation of trastuzumab

Exclusion Criteria:
- Evidence of ischemic heart disease as determined by the study cardiologist
- Significant valvular heart disease (aortic stenosis [AS] with aortic valve area [AVA] < 1.5 and severe aortic regurgitation [AR] and mitral regurgitation [MR])
- History of familial cardiomyopathy
- Recent documented myocarditis within 2 months of consent

- History of infiltrative cardiomyopathy or restrictive cardiomyopathy
- Epidermal growth factor receptor (eGFR) <50 by Mayo or Cockcroft formula
- Liver function tests >3 × upper limit of normal
- NYHA class IV heart failure
- Inotropic dependence
- Unstable or life-threatening arrhythmia
- Coagulopathy international normalized ratio (INR) >1.5
- Mechanical or bioprosthetic heart valve
- Cardiogenic shock
- Breastfeeding and/or pregnant women
- Autoimmune disorders on current immunosuppressive therapy
- Active infection not responding to appropriate therapy as determined by the study chair
- Trastuzumab treatment within the last 3 months

Chapter 2
Donor Bone Marrow-Derived Mesenchymal Stem Cells in Controlling Heart Failure in Patients with Cardiomyopathy Caused by Anthracyclines

Brief Summary:

This randomized pilot phase I trial studies the side effects of donor bone marrow-derived mesenchymal stem cells in controlling heart failure in patients with cardiomyopathy caused by anthracyclines. Donor bone marrow-derived mesenchymal stem cells may help control symptoms of heart failure and improve heart function.

Condition or disease	Intervention/treatment	Phase
Cardiomyopathy Heart failure	Other: Best practice Other: Laboratory biomarker analysis Biological: Mesenchymal stem cell transplantation	Phase 1

Primary Objective:

To demonstrate the safety of allogeneic human mesenchymal stem cells (hMSCs) administered via intravenous infusion and transendocardial injection in patients with left ventricular (LV) dysfunction and heart failure secondary to chemotherapy with anthracyclines.

Secondary Objective:

To demonstrate the efficacy of allogeneic hMSCs administered via intravenous infusion and transendocardial injection in patients with left ventricular dysfunction (left ventricular ejection fraction [LVEF] <40%) and heart failure secondary to treatment with anthracyclines.

Outline: Patients are randomized to 1 of 3 arms.

Arm I: Patients receive hMSCs intravenously (IV) over 10–20 minutes on days 1, 14, 21, and 28 and standard-of-care treatment for heart failure in the absence of disease progression or unacceptable toxicity.

MD Anderson Cancer Center, Houston, Texas, United States
ClinicalTrials.gov Identifier: NCT02962661

J. N. Weiss, *Stem Cell Surgery Trials in Heart Failure and Diabetes*, https://doi.org/10.1007/978-3-030-78010-4_2

Arm II: Patients receive hMSCs transendocardially for a total of 15 injections and standard-of-care treatment for heart failure in the absence of disease progression or unacceptable toxicity.

Arm III: Patients receive standard-of-care treatment for heart failure.

After the completion of study treatment, patients are followed up periodically.

Study Type:	Interventional (Clinical Trial)
Estimated Enrollment:	72 participants
Allocation:	Randomized
Intervention Model:	Parallel Assignment
Masking:	None (Open Label)
Primary Purpose:	Treatment
Official Title:	Randomized 3-Arm Trial with Standard of Care Alone vs Either Intravenous Infusion or Transendocardial Injection of Allogeneic Bone Marrow-Derived Multipotent Mesenchymal Stem Cells (MSCs) Plus Standard of Care in Patients with Anthracycline-Associated Cardiomyopathy
Actual Study Start Date:	July 18, 2020
Estimated Primary Completion Date:	July 30, 2023
Estimated Study Completion Date:	July 30, 2023

Arm	Intervention/treatment

Experimental: Arm I (hMSCs IV)
Patients receive hMSCs IV over 10–20 minutes on days 1, 14, 21, and 28 and standard-of-care treatment for heart failure in the absence of disease progression or unacceptable toxicity
	Other: Best practice
	Given standard of care
	Other names:
	    Standard of care
	    Standard therapy
	Other: Laboratory biomarker analysis correlative studies
	Biological: Mesenchymal stem cell transplantation intravenous infusion (IV)

Experimental: Arm II (hMSCs transendocardially)
Patients receive hMSCs transendocardially for a total of 15 injections and standard-of-care treatment for heart failure in the absence of disease progression or unacceptable toxicity
	Other: Best practice
	Given standard of care
	Other names:
	    Standard of care
	    Standard therapy
	Other: Laboratory biomarker analysis correlative studies
	Biological: Mesenchymal stem cell transplantation transendocardially (injection)

Arm	Intervention/treatment
Active comparator: Arm III (standard of care) Patients receive standard-of-care treatment for heart failure	Other: Best practice Given standard of care Other names: Standard of care Standard therapy Other: Laboratory biomarker analysis correlative studies

Primary Outcome Measures:

1. Incidence of adverse events [time frame: up to 6 months]

   Statistical analyses of safety will be descriptive.

2. Change in the left ventricular ejection fraction (LVEF) [time frame: baseline to 6 months]

   The comparison will be between the two groups of patients.

Secondary Outcome Measures:

1. Change in the improvement of left ventricular (LV) systolic function as assessed by LVEF [time frame: baseline up to 6 months]

   As regards statistical analyses, the results of the trial will be displayed in table format. Confidence intervals of the differences in the change from baseline between each investigational group and the control group will be provided. If both investigational groups are significant at the $P < .05$ level, then the two investigational drugs can be compared using a gatekeeping procedure. These intervals and the associated P-values will be calculated using the two-sample t-tests, with no adjustments for multiple comparisons.

2. LV end-systolic and end-diastolic volumes as determined by contrast-enhanced 2-dimensional(D)/3D echography [time frame: up to 6 months]

   As regards statistical analyses, the results of the trial will be displayed in table format. Confidence intervals of the differences in the change from baseline between each investigational group and the control group will be provided. If both investigational groups are significant at the $P < .05$ level, then the two investigational drugs can be compared using a gatekeeping procedure. These intervals and the associated P-values will be calculated using the two-sample t-tests, with no adjustments for multiple comparisons.

3. Cardiac death [time frame: up to 6 months]

   As regards statistical analyses, the results of the trial will be displayed in table format. Confidence intervals of the differences in the change from baseline between each investigational group and the control group will be provided. If both investigational groups are significant at the $P < .05$ level, then the two investigational drugs can be compared using a gatekeeping procedure. These intervals and the associated P-values will be calculated using the two-sample t-tests, with no adjustments for multiple comparisons.

4. Re-hospitalization after heart failure [time frame: up to 6 months]

As regards statistical analyses, the results of the trial will be displayed in table format. Confidence intervals of the differences in the change from baseline between each investigational group and the control group will be provided. If both investigational groups are significant at the $P < .05$ level, then the two investigational drugs can be compared using a gatekeeping procedure. These intervals and the associated P-values will be calculated using the two-sample t-tests, with no adjustments for multiple comparisons.

5. Aborted death from an automatic implantable cardioverter defibrillator (AICD) firing [time frame: up to 6 months]

As regards statistical analyses, the results of the trial will be displayed in table format. Confidence intervals of the differences in the change from baseline between each investigational group and the control group will be provided. If both investigational groups are significant at the $P < .05$ level, then the two investigational drugs can be compared using a gatekeeping procedure. These intervals and the associated P-values will be calculated using the two-sample t-tests, with no adjustments for multiple comparisons.

6. Nonfatal myocardial infarction [time frame: up to 6 months]

As regards statistical analyses, the results of the trial will be displayed in table format. Confidence intervals of the differences in the change from baseline between each investigational group and the control group will be provided. If both investigational groups are significant at the $P < .05$ level, then the two investigational drugs can be compared using a gatekeeping procedure. These intervals and the associated P-values will be calculated using the two-sample t-tests, with no adjustments for multiple comparisons.

7. Revascularization [time frame: up to 6 months]

As regards statistical analyses, the results of the trial will be displayed in table format. Confidence intervals of the differences in the change from baseline between each investigational group and the control group will be provided. If both investigational groups are significant at the $P < .05$ level, then the two investigational drugs can be compared using a gatekeeping procedure. These intervals and the associated P-values will be calculated using the two-sample t-tests, with no adjustments for multiple comparisons.

Eligibility

Ages Eligible for Study:	18 Years to 80 Years (Adult, Older Adult)
Sexes Eligible for Study:	All
Accepts Healthy Volunteers:	No

Inclusion Criteria:
- Patients with LVEF = <40% documented from treatment with anthracyclines for any malignancy at any dose at any time without evidence of other causes of cardiomyopathy
- Documented New York Heart Association (NYHA) classes I, II, and III
- Patients with persistent LV dysfunction 90 days after discontinuation of trastuzumab
- Able to perform the 6-minute walk test
- Been treated with appropriate maximal medical therapy for heart failure
- Patient or legally authorized representative able to sign informed consent

Exclusion Criteria:
- Evidence of ischemic heart disease as determined by the study cardiologist
- Significant valvular heart disease (aortic stenosis [AS] with aortic valve area [AVA] < 1.5 and severe aortic regurgitation [AR] and mitral regurgitation [MR])
- History of familial cardiomyopathy
- Recent documented myocarditis within 2 months of enrollment
- History of infiltrative cardiomyopathy or restrictive cardiomyopathy
- Estimated glomerular filtration rate (eGFR) <50 by Mayo or Cockcroft formula
- Presence of left ventricular thrombus as documented by echocardiography or left ventriculogram
- Liver function tests >3 × the upper limit of normal
- NYHA class IV heart failure
- Inotropic dependence
- Unstable or life-threatening arrhythmia
- Coagulopathy international normalized ratio (INR) >1.5
- Mechanical or bioprosthetic heart valve
- Cardiogenic shock
- Breastfeeding and/or pregnant women
- Autoimmune disorders on current immunosuppressive therapy
- Active infection not responding to appropriate therapy as determined by the study chair
- Trastuzumab treatment within the last 3 months
- Automatic implantable cardioverter defibrillator (AICD) placement within the last 30 days
- AICD firing within the last 30 days

Chapter 3
The Transendocardial Autologous Cells (hMSC) or (hMSC) and (hCSC) in Ischemic Heart Failure Trial

Brief Summary:

Before initiating the full randomized study, a pilot safety phase will be performed. In this phase the composition of cells administered via the Biosense Webster MyoStar NOGA Injection Catheter System will be tested. The randomized portion of the study will be conducted after a full review of the safety data from the pilot phase by the data safety monitoring board.

Following the pilot phase of five (5), fifty (50) patients scheduled to undergo cardiac catheterization and meeting all inclusion/exclusion criteria will be evaluated at baseline.

Patients will be randomized in a 2:2:1 ratio to one of three treatment strategies.

Condition or disease	Intervention/treatment	Phase
Chronic ischemic left Ventricular dysfunction myocardial infarction	Drug: Autologous hMSCs Drug: Autologous Human C-Kit CSCs II Drug: Placebo Device: Biosense Webster MyoStar NOGA Injection Catheter System	Phase 1 Phase 2

Detailed Description:

A phase I/II, randomized, placebo-controlled study of the safety and efficacy of transendocardial injection of autologous human cells (mesenchymal or the combination of MSC and cardiac stem cells) in patients with chronic ischemic left ventricular dysfunction and heart failure secondary to myocardial infarction.

A total of 55 subjects are included in the study, 5 in the pilot phase and 50 in the randomized phase.

ISCI/University of Miami, Miami, Florida, United States
ClinicalTrials.gov Identifier: NCT02503280

J. N. Weiss, *Stem Cell Surgery Trials in Heart Failure and Diabetes*, https://doi.org/10.1007/978-3-030-78010-4_3

Patients with chronic ischemic left ventricular dysfunction and heart failure secondary to MI scheduled to undergo cardiac catheterization.

Study Type:	Interventional (Clinical Trial)
Estimated Enrollment:	55 Participants
Allocation:	Randomized
Intervention Model:	Parallel Assignment
Masking:	Single (Participant)
Primary Purpose:	Treatment
Official Title:	A Phase I/II, Randomized, Placebo-Controlled Study of the Safety and Efficacy of Transendocardial Injection of Autologous Human Cells (Mesenchymal or the Combination of MSC and Cardiac Stem Cells) in Patients with Chronic Ischemic Left Ventricular Dysfunction and Heart Failure Secondary to Myocardial Infarction.
Estimated Study Start Date:	March 1, 2025
Estimated Primary Completion Date:	March 2030
Estimated Study Completion Date:	March 2032

Arm	Intervention/treatment
Experimental: Group A – Autologous hMSCs Autologous hMSCs: 40 million cells/mL delivered in 0.5-mL injection volumes times 10 injections for a total of 2×10^8 (200 million) hMSCs. The Biosense Webster MyoStar NOGA Injection Catheter System will be used in the delivery of the study drug	Drug: Autologous hMSCs Autologous hMSCs: 40 million cells/mL delivered in 0.5-mL injection volumes times 10 injections for a total of 2×10^8 (200 million) hMSCs Other name: Autologous human mesenchymal stem cells (hMSCs) Device: The Biosense Webster MyoStar NOGA Injection Catheter System will be used to administer the study drug Other name: NOGA
Experimental: Group B – Autologous Human C-Kit CSCs II Autologous hMSCs PLUS autologous C-Kit hCSCs: Mixture of 39.8 million hMSCs and 0.2 million C-Kit hCSCs/mL delivered in 0.5-mL injection volumes times 10 injections for a total of 1.99×10^8 (199 million) hMSCs and 1 million C-Kit hCSCs. The Biosense Webster MyoStar NOGA Injection Catheter System will be used in the delivery of the study drug	Drug: Autologous Human C-Kit CSCs II Autologous hMSCs PLUS autologous C-Kit hCSCs: Mixture of 39.8 million hMSCs and 0.2 million C-Kit hCSC/mL delivered in 0.5-mL injection volumes times 10 injections for a total of 1.99×10^8 (199 million) hMSCs and 1 million C-Kit hCSCs Other name: Autologous Human C-Kit Cardiac Stem Cells (CSCs) II Device: The Biosense Webster MyoStar NOGA Injection Catheter System will be used to administer the study drug Other name: NOGA

Arm	Intervention/treatment
Placebo comparator: Placebo Placebo (ten 0.5-mL injections of phosphate-buffered saline [PBS] and 1% human serum albumin [HSA]). The Biosense Webster MyoStar NOGA Injection Catheter System will be used in the delivery of the study drug	Drug: Placebo Placebo (ten 0.5-mL injections of phosphate-buffered saline [PBS] and 1% human serum albumin [HSA]) Device: The Biosense Webster MyoStar NOGA Injection Catheter System System will be used to administer the study drug Other name: NOGA

Primary Outcome Measures:
1. Incidence of any treatment emergent serious adverse events (TE-SAEs) [time frame: 1 month post-catheterization]

   Incidence (at 1 month post-catheterization) of any treatment-emergent serious adverse events (TE-SAEs), defined as the composite of death, non-fatal MI, stroke, hospitalization for worsening heart failure, cardiac perforation, pericardial tamponade, sustained ventricular arrhythmias (characterized by ventricular arrhythmias lasting longer than 15 seconds or with hemodynamic compromise), or atrial fibrillation.

Secondary Outcome Measures:
1. Treatment emergent adverse event rates [time frame: at the 6-month and 12-month visits]

   Rate of adverse events occurring

2. Ectopic tissue formation [time frame: at the 6-month and 12-month visits]

   Ectopic tissue formation (as identified from the MRI scans of the chest, abdomen, and pelvis)

3. 48-hour ambulatory electrocardiogram (ECG) recordings [time frame: at the 6-month and 12-month visits]

   Electrocardiogram (ECG) recordings measured over 48 hours

4. Hematology value changes post-catheterization [time frame: at the 6-month and 12-month visits]

   Hematology value changes will be observed at the 6-month and 12-month visits post-catheterization.

5. Urinalysis result changes post-catheterization [time frame: at the 6-month and 12-month visits]

   Urinalysis results changes will be observed at the 6-month and 12-month visits post-catheterization.

6. Clinical chemistry values post-catheterization [time frame: at the 6-month and 12-month visits]

Clinical chemistry value changes will be observed at the 6-month and 12-month visits post-catheterization.

7. Pulmonary function [time frame: at the 6-month and 12-month visits]

Pulmonary function – forced expiratory volume in 1 second (FEV1) results

8. Serial troponin I values [time frame: every 12 hours for the first 48 hours post-cardiac catheterization]

Serial troponin I values (every 12 hours for the first 48 hours post-cardiac catheterization)

9. Creatine kinase-MB (CK-MB) value changes post-catheterization [time frame: every 12 hours for the first 48 hours post-cardiac catheterization]

CK-MB values (every 12 hours for the first 48 hours post-cardiac catheterization)

10. Post-cardiac catheterization echocardiogram [time frame: Day 1 post-echocardiogram]

Echocardiogram performed after cardiac catheterization

11. Magnetic resonance imaging (MRI) measures of infarct scar size (ISS) [time frame: at the 6-month and 12-month visits]

Document infarct scar size (ISS) via magnetic resonance imaging (MRI).

12. Echocardiographic measures of infarct scar size (ISS) [time frame: at the 6-month and 12-month visits]

Document infarct scar size (ISS) via echocardiographic procedure.

13. Magnetic resonance imaging (MRI) of the left regional ventricular function [time frame: at the 6-month and 12-month visits]

Document left regional ventricular function via magnetic resonance imaging (MRI).

14. Echocardiographic measures of the left regional ventricular function [time frame: at the 6-month and 12-month visits]

Document left regional ventricular function via echocardiographic procedure.

15. Magnetic resonance imaging (MRI) of the global ventricular function [time frame: at the 6-month and 12-month visits]

Document global ventricular function via magnetic resonance imaging (MRI).

16. Echocardiographic measures of global ventricular function [time frame: at the 6-month and 12-month visits]

Document global ventricular function via echocardiographic procedure.

17. Tissue perfusion measured by MRI [time frame: at the 6-month and 12-month visits]

Measure tissue perfusion via magnetic resonance imaging (MRI).

18. Peak oxygen consumption (Peak VO2) (by treadmill determination) [time frame: at the 6-month and 12-month visits]

Peak VO2 oxygen consumption determined by utilizing treadmill.

19. Six-minute walk test [time frame: at the 6-month and 12-month visits]

Evaluate functional capacity via the six-minute walk test.

20. New York Heart Association (NYHA) functional class [time frame: at the 6-month and 12-month visits]

Evaluate functional capacity via New York Heart Association (NYHA) class determination.

21. Minnesota Living with Heart Failure (MLHF) questionnaire [time frame: at the 6-month and 12-month visits]

Evaluate quality of life changes using the Minnesota Living with Heart Failure (MLHF) Questionnaire.

22. Incidence of major adverse cardiac events (MACE) [time frame: at the 6-month and 12-month visits]

Incidence of major adverse cardiac events (MACE), defined as the composite incidence of (1) death, (2) hospitalization for worsening HF, or (3) non-fatal recurrent MI

Ages Eligible for Study:	21 Years to 89 Years (Adult, Older Adult)
Sexes Eligible for Study:	All
Accepts Healthy Volunteers:	No

Inclusion Criteria:
- In order to participate in this study, a patient MUST:

  1. Be ≥21 and <90 years of age.
  2. Provide written informed consent.
  3. Have a diagnosis of chronic ischemic left ventricular dysfunction secondary to myocardial infarction (MI) as defined by the following: screening MRI must show an area of akinesis, dyskinesis, or severe hypokinesis associated with evidence of myocardial scarring based on delayed hyperenhancement following gadolinium infusion.

4. Been treated with appropriate maximal medical therapy for heart failure or post-infarction left ventricular dysfunction. For beta-blockade, the patient must have been on a stable dose of a clinically appropriate beta-blocker for 3 months. For angiotensin-converting enzyme inhibition, the patient must have been on a stable dose of a clinically appropriate agent for 1 month.
5. Be a candidate for cardiac catheterization.
6. Have an ejection fraction $\leq$50% by gated blood pool scan, two-dimensional echocardiogram, cardiac MRI, or left ventriculogram within the prior 6 months and not in the setting of a recent ischemic event.

Exclusion Criteria:
- In order to participate in this study, a patient MUST NOT:

1. Have a baseline glomerular filtration rate <50 mL/min/1.73 m^2.
2. Have a known, serious radiographic contrast allergy.
3. Have a mechanical aortic valve or heart constrictive device.
4. Have a documented presence of aortic stenosis (aortic stenosis graded as $\geq$+2 equivalent to an orifice area of 1.5 cm^2 or less).
5. Have a documented presence of moderate to severe aortic insufficiency (echocardiographic assessment of aortic insufficiency graded as $\geq$+2).
6. Require coronary artery revascularization. Patients who require or undergo revascularization procedures should undergo these procedures for a minimum of 3 months in advance of treatment within this study. In addition, patients who develop a need for revascularization following enrollment will be submitted for this therapy without delay.
7. Evidence of a life-threatening arrhythmia (nonsustained ventricular tachycardia $\geq$20 consecutive beats or complete heart block) or QTc interval > 550 ms on screening ECG.
8. AICD firing in the past 60 days prior to the procedure.
9. Have unstable angina within 2 weeks of the planned procedure.
10. Have a hematologic abnormality as evidenced by hematocrit <25%, white blood cell <2500/uL, or platelet values <100,000/uL without another explanation.
11. Have liver dysfunction, as evidenced by enzymes (AST and ALT) greater than three times the ULN.
12. Have a coagulopathy = (INR > 1.3) not due to a reversible cause (i.e., Coumadin). Patients on Coumadin will be withdrawn 5 days before the procedure and confirmed to have an INR < 1.3. Patients who cannot be withdrawn from Coumadin will be excluded from enrollment.
13. Have known allergies to penicillin or streptomycin.
14. Have a contraindication to performance of an MRI scan.
15. Be an organ transplant recipient.
16. Have a clinical history of malignancy within 5 years (i.e., patients with prior malignancy must be disease-free for 5 years), except curatively treated basal cell carcinoma, squamous cell carcinoma, or cervical carcinoma.

17. Have a non-cardiac condition that limits lifespan to <1 year.
18. Have a history of drug or alcohol abuse within the past 24 months.
19. Be on chronic therapy with an immunosuppressant medication, such as corticosteroids or TNFα antagonists.
20. Be serum-positive for HIV, hepatitis BsAg, or hepatitis C.
21. Be currently participating (or participated within the previous 30 days) in an investigational therapeutic or device trial.
22. Be a female who is pregnant, nursing, or of childbearing potential while not practicing effective contraceptive methods. Female patients must undergo a blood or urine pregnancy test at screening and within 36 hours prior to injection.

Chapter 4
Regression of Fibrosis and Reversal of Diastolic Dysfunction in HFPEF Patients Treated with Allogeneic CDCs

Brief Summary:

Perform a randomized, double-blind, placebo-controlled Phase 2a feasibility study to determine whether treatment of HFPEF patients with intracoronary allogeneic CDCs affects clinical functional status (QOL scores), exercise tolerance (6MHW), exercise hemodynamics (supine exercise ergometry during right heart catheterization), myocardial interstitial fibrosis (MRI with native T1 mapping and calculation of extracellular volume [ECV] after gadolinium administration), macroscopic fibrosis by delayed gadolinium enhancement (DGE), and diastolic function (catheterization, echocardiography, BNP).

Treatment of patients with symptomatic hypertensive heart disease-induced HFPEF with allogeneic CDCs will be safe and will improve clinical functional status, exercise tolerance/hemodynamics, myocardial interstitial structure, and diastolic function; the mechanisms underlying these improvements will be reflected in the changes in plasma biomarkers that indicate a reduction in pro-inflammatory and pro-fibrotic signaling.

Condition or disease	Intervention/treatment	Phase
Congestive heart failure	Biological: Allogeneic derived cells Biological: Placebo/control arm	Phase 2

Medical University of South Carolina, Charleston, South Carolina, United States.
ClinicalTrials.gov Identifier: NCT02941705.

© The Author(s), under exclusive license to Springer Nature   21
Switzerland AG 2022
J. N. Weiss, *Stem Cell Surgery Trials in Heart Failure and Diabetes*,
https://doi.org/10.1007/978-3-030-78010-4_4

Arm	Intervention/treatment
Active comparator: RECEIVED CELLS This arm will receive the CDC/CAP 1002 solution	Biological: Allogeneic derived cell Patients will have the CAP-1002 solution delivered through a coronary catheter inserted into the right and left coronary arteries using standard techniques in the cardiac catheterization laboratory. A right heart catheter will be used to obtain baseline (pre-infusion) hemodynamics Other name: CAP-1002
Placebo comparator: CONTROL ARM This arm will receive a solution during randomization but will not receive the CDCs	Biological: Placebo/control arm Patients will receive the placebo through a coronary catheter inserted into the right and left coronary arteries using standard techniques in the cardiac catheterization laboratory. A right heart catheter will be used to obtain baseline (pre-infusion) hemodynamics Other name: Placebo

Primary Outcome Measures:
1. The safety profile of CAP 1002; any subjects experiencing any safety-related events during or post intracoronary delivery and during the follow-up period [time frame: 3 years].

Any subjects experiencing any safety-related events during or post intracoronary delivery and during the follow-up period.

Safety outcomes will be measured through TIMI flow 0–2, acute myocarditis within 1 month of intracoronary infusion, ventricular tachycardia or ventricular fibrillation within 72 hours of intracoronary infusion, sudden unexpected death within 72 hours of intracoronary infusion defined as occurring 1 hour within symptom onset or witnessed death in a subject previously observed to be well within the preceding 24 hours without an identified cause, or a major adverse cardiac event within 72 hours of intracoronary infusion.

Eligibility Criteria

Ages eligible for study:	50 Years and older (Adult, Older Adult)
Sexes eligible for study:	All
Accepts healthy volunteers:	No

Inclusion Criteria:
1. ≥ 50 years old, male or female
2. LVEF $\geq 50\%$.
3. Symptoms and physical findings of chronic heart failure (NYHA class II – ambulatory IV).
4. Treatment with a stable, maximally tolerated dose of diuretic(s) for a minimum of 30 days prior to randomization.

5. Left atrial (LA) enlargement defined by at least one of the following: LA width (diameter) ≥3.8 cm or LA length ≥5.0 cm or LA area ≥20 cm² or LA volume ≥55 mL or LA volume index ≥29 mL/m².
6. BNP >125 pg/mL for patients in NSR or >150 pg/mL for patients in AF (BNP are BMI-corrected) or resting PCWP >15 mmHg, or exercise PCWP >18 mmHg.

Exclusion Criteria Specific to HFPEF:
1. Any prior echocardiographic measurement of LVEF <40%.
2. Acute coronary syndrome (including MI), cardiac surgery, other major CV surgery, or percutaneous coronary intervention (PCI) within 3 months prior to randomization.
3. Unrevascularized, hemodynamically significant CAD (FFR < 0.75).
4. Current acute decompensated HF.
5. Alternative diagnoses that, in the opinion of the investigator, could account for the patient's HF symptoms (i.e., dyspnea, fatigue), such as severe pulmonary disease (i.e., requiring home oxygen, chronic nebulizer therapy, chronic oral steroid therapy), hemoglobin (Hgb) < 10 g/dL, and body mass index (BMI) > 40 kg/m².
6. Use of investigational drugs or treatments at the time of enrollment.
7. Systolic blood pressure > 150 mmHg but <180 mmHg unless receiving 3 or more antihypertensive drugs.
8. History of any dilated cardiomyopathy; right-sided HF in the absence of left-sided structural heart disease; pericardial constriction, genetic hypertrophic cardiomyopathy, or infiltrative cardiomyopathy; clinically significant congenital heart disease; hemodynamically significant valvular heart disease.
9. Stroke, transient ischemic attack, carotid surgery, or carotid angioplasty within the 3 months.
10. Uncontrolled dysrhythmia; symptomatic or sustained ventricular tachycardia or atrial fibrillation or flutter with a resting ventricular rate > 110 beats per minute (bpm).
11. Prior major organ transplant or intent to transplant (i.e., on transplant list).
12. Hepatic disease as determined by any one of the following: SGOT (AST) or SGPT (ALT) values exceeding 3× the upper limit of normal (ULN), bilirubin >1.5 mg/dL; history of chronic viral hepatitis.
13. Chronic kidney disease with eGFR <30 mL/min/1.73 m²; serum potassium >5.5 mmol/L (mEq/L).
14. History or presence of any other disease with a life expectancy of <3 years.
15. Non-compliance with medical regimens.
16. Drug or alcohol abuse within the last 12 months.
17. History of malignancy within the past 5 years.
18. Pregnant or nursing (lactating) women confirmed by a positive human chorionic gonadotropin (hCG); women of childbearing potential (physiologically capable of becoming pregnant), unless using highly effective contraception methods during the study period.

Exclusion Criteria Specific to CAP-1002 (Not Listed Above):

1. Diagnosis of active myocarditis.
2. Known hypersensitivity to contrast agents or previous H/O HIT.
3. Known hypersensitivity to dimethyl sulfoxide (DMSO).
4. Known hypersensitivity to bovine products.
5. Active infection not responsive to treatment.
6. Active allergic reactions, connective tissue diseases, or autoimmune disorders.
7. History of cardiac tumor or cardiac tumor demonstrated on screening.
8. History of previous stem cell therapy.
9. History of treatment with immunosuppressive agents, including chronic systemic corticosteroids, biologic agents targeting the immune system, anti-tumor and anti-neoplastic drugs, or anti-VEGF within 6 months prior to enrollment (excluding drug-eluting coronary stents).
10. Human immunodeficiency virus (HIV) infection.

Chapter 5
Efficacy and Safety of Allogeneic Mesenchymal Precursor Cells (Rexlemestrocel-L) for the Treatment of Heart Failure

Brief Summary:

The primary objective of this study is to determine whether transendocardial delivery of allogeneic human bone marrow-derived MPCs (rexlemestrocel-L) is effective in the treatment of chronic heart failure due to LV systolic dysfunction.

Condition or disease	Intervention/treatment	Phase
Chronic heart failure	Biological: Allogeneic mesenchymal precursor cells (MPC) Other: Sham comparator	Phase 3

Detailed Description:

The purpose of this study is to evaluate the efficacy and safety of a single transendocardial delivery in the cardiac catheterization laboratory of human bone marrow-derived allogeneic MPCs (rexlemestrocel-L) for improving clinical outcomes (HF-MACE), preventing further adverse cardiac remodeling (LVESV and LVEDV), and increasing exercise capacity (6MWT) in patients with chronic HF due to LV systolic dysfunction of either ischemic or nonischemic etiology who have received optimal medical/revascularization therapy.

Mesoblast Investigational Site 10757 – Cardiology, P.C., Birmingham, Alabama, United States
Mesoblast Investigational Site 13262 – University of Alabama at Birmingham Hospital, Birmingham, Alabama, United States
Mesoblast Investigational Site 10779 – Mercy Gilbert Medical Center, Gilbert, Arizona, United States
(and 56 more...)
ClinicalTrials.gov Identifier: NCT02032004.

Study Type:	Interventional (Clinical Trial)
Actual enrollment:	566 participants
Allocation:	Randomized
Intervention model:	Parallel assignment
Masking:	Triple (participant, investigator, outcomes assessor)
Primary purpose:	Treatment
Official title:	Double-blind, randomized, sham procedure-controlled, Parallel-group efficacy and safety study of allogeneic Mesenchymal precursor cells (Rexlemestrocel-L) in chronic heart failure due to LV systolic dysfunction (ischemic or nonischemic) dream HF-1
Actual study start date:	February 14, 2014
Estimated primary completion date:	May 29, 2020
Estimated study completion date:	May 29, 2020

Arm	Intervention/treatment

Experimental: Allogeneic mesenchymal precursor cells
Participants randomly assigned to treatment will undergo a single index cardiac catheterization involving transendocardial delivery of rexlemestrocel-L into the myocardium at a cell injection center by an interventional cardiology team not involved in the review or assessment of subsequent study results
	Biological: Allogeneic mesenchymal precursor cells (MPC)
	Rexlemestrocel-L consists of human bone marrow-derived allogeneic MPCs isolated from bone mononuclear cells with anti-STRO-3 antibodies, expanded ex vivo, and cryopreserved
	Other names:
	    MPC.
	    Rexlemestrocel-L.

Sham comparator: Control treatment Participants randomly assigned to control treatment will undergo a single cardiac catheterization involving a scripted sham cardiac mapping and cell delivery procedure at a cell injection center by an interventional cardiology team not involved in the review or assessment of subsequent study results	Other: Sham comparator The sham procedure will be staged to script and will not include an actual cardiac mapping or delivery of rexlemestrocel-L

Primary Outcome Measures:
1. Time-to-recurrent non-fatal decompensated heart failure major adverse cardiac events (HF-MACE) that occur prior to the first terminal cardiac event (TCE) [time frame: 6 months minimum].

Secondary Outcome Measures:
1. Time-to-first terminal cardiac event (TCE) [time frame: 6 months minimum].
2. Time-to-hospital admissions for non-fatal decompensated HF events [time frame: 6 months minimum].

3. Time-to-urgent care outpatient HF visits [time frame: 6 months minimum].
4. Time-to-successfully resuscitated cardiac death (RCD) events [time frame: 6 months minimum].
5. Total length of in-hospital stay in intensive care unit for non-fatal decompensated HF events [time frame: 6 months minimum].
6. Time-to-first HF-MACE (composite of hospital admissions for decompensated HF, urgent care outpatient HF visit, and successful RCD events) [time frame: 6 months minimum].
7. Time-to-first HF-MACE (composite of hospital admissions for decompensated HF, urgent care outpatient HF visit, successful RCD events or TCE) [time frame: 6 months minimum].
8. Time-to-cardiac death [time frame: 6 months minimum].
9. Time-to-all-cause death [time frame: 6 months minimum].
10. Time-to-first non-fatal MI (myocardial infarction), non-fatal CVA (cerebrovascular attack), or coronary artery revascularization [time frame: 6 months minimum].
11. Left ventricular (LV) remodeling in LVESV determined by 2D echocardiography [time frame: screening; day 0 (post-procedure); months 3, 6, and 12; and every 12 months thereafter during the study period (6 months minimum)].
12. Correlations between baseline LVESV <= 100 mL and LVESV >100 mL and clinical outcomes [time frame: 6 months minimum].
13. Correlations between baseline LVESV <= 100 mL and LVESV >100 mL and change in month 6 to baseline LVESV and clinical outcomes [time frame: 6 months minimum].
14. LV remodeling in LVEDV determined by 2D echocardiography [time frame: screening; day 0 (post-procedure); months 3, 6, and 12; and every 12 months thereafter during the study period (6 months minimum)].
15. Overall left ventricular systolic performance as assessed by LVEF (RVG or echocardiogram) [time frame: screening; day 0 (post-procedure); months 3, 6, and 12; and every 12 months thereafter during the study period (6 months minimum)].
16. Functional exercise capacity as assessed by the 6-minute walk test [time frame: screening; months 3, 6, and 12; and every 12 months thereafter during the study period (6 months minimum)].
17. Functional status by the New York Heart Association (NYHA) class [time frame: screening; months 3, 6, and 12; and every 12 months thereafter during the study period (6 months minimum)].
18. Quality of Life Measure – Minnesota Living with Heart Failure (MLHF) questionnaire [time frame: screening; months 3, 6, and 12; and every 12 months thereafter during the study period (6 months minimum)].
19. Quality of Life Measure – EuroQoL 5-dimensional (EQ-5D) questionnaire [time frame: screening; months 3, 6, and 12; and every 12 months thereafter during the study period (6 months minimum)].

20. Safety as assessed by the occurrence of adverse events related to the index cardiac catheterization on day 0 [time frame: day 0 through discharge from day 0 hospitalization].
21. Safety as assessed by the occurrence of treatment-emergent adverse events [time frame: screening through 6 months minimum].
22. Safety as assessed by clinical laboratory tests (serum chemistry [ALT, AST, alkaline phosphate, GGT, LDH, BUN, creatinine, uric acid, total bilirubin] and hematology [hematocrit, hemoglobin, WBC, eosinophils, ANC, platelet count]) [time frame: screening; day 0 (post-procedure); day 10; months 1, 3, 6, and 12; and every 6 months thereafter during the study period (6 months minimum)].
23. Safety as assessed by urinalysis (blood, glucose, ketones, total protein) [time frame: screening; day 10; months 1, 3, 6, and 12; and every 6 months thereafter during the study period (6 months minimum)].
24. Safety as assessed by vital signs (pulse, systolic BP, diastolic BP) [time frame: screening; day 0 (pre- and post-procedure); day 1; day 10; months 1, 3, 6, and 12; and every 6 months thereafter during the study period (6 months minimum)].
25. Safety as assessed by 12-lead electrocardiogram (ECG) findings (QTcF, QTcB, QT, QRS complex, HR, and T waves) [time frame: screening; day 0 (pre- and post-procedure); day 1; day 10; months 1, 3, 6, and 12; and every 6 months thereafter during the study period (6 months minimum)].
26. Safety as assessed by telemetry monitoring findings (clinically significant arrhythmias) [time frame: day 0 through day 0 overnight post-procedure].
27. Safety as assessed by rhythm analysis (specifically, ventricular arrhythmias) by interrogation of any implanted device capable of defibrillation [time frame: day 10; months 1, 3, 6, and 12; and every 6 months thereafter during the study period (6 months minimum)].
28. Safety as assessed by 24-hour Holter monitoring (HR, rate and duration of arrhythmias, a-fib average rate, supra- and ventricular ectopy singles/couplets/runs/totals, sustained and non-sustained ventricular tachycardia, longest pause RR duration, total pauses) [time frame: screening, day 0 (post-procedure), day 10, months 1 and 3].
29. Safety as assessed by physical examination findings judged as clinically significant changes from baseline by the investigator or newly occurring abnormalities (including weight) [time frame: screening, month 12 and every 12 months thereafter until study conclusion (weight measured at screening; day 0; pre- and post-procedure; day 1; day 10; months 1, 3, 6, and 12; and every 6 months thereafter)].
30. Safety as assessed by important cardiovascular events from adjudicated data [time frame: 6 months minimum].

Other Outcome Measures:
1. Pharmacodynamics measures (NT-proBNP and hsCRP) [time frame: screening; months 3, 6, and 12; and every 12 months thereafter until the study conclusion].
2. Pharmacogenomics (PGx) analysis [time frame: screening (only from those subjects who provide consent to participate in PGx sample collection)].
3. Immunogenicity measures (panel reactive antibodies, donor-specific antibody, if PRA test is positive, antibodies against bovine and murine products) [time frame: screening; day 10; months 1, 3, 6, and 12].

Ages eligible for study:	18 years to 80 years (adult, older adult)
Sexes eligible for study:	All
Accepts healthy volunteers:	No

Inclusion Criteria:
- The patient is 18 to 80 years of age, inclusive; both men and women will be enrolled.
- The patient has a diagnosis of chronic HF of ischemic or nonischemic etiology for at least 6 months.
- The patient is on stable, optimally tolerated dosages of HF therapies including beta-blockers (approved for country-specific usage), angiotensin-converting enzyme (ACE) inhibitors or angiotensin-receptor blockers (ARBs), and/or aldosterone antagonists, without change in dose for at least 1 month before study intervention.
- The patient is on a stable, outpatient, oral diuretic dosing regimen in which the patient remains clinically stable during screening.
- If other criteria apply, please contact the investigator.

Exclusion Criteria:
- The patient has NYHA Functional Class I or Functional Class IV symptoms.
- If other criteria apply, please contact the investigator.

Publications:
Borow KM, Yaroshinsky A, Greenberg B, Perin EC. Phase 3 DREAM-HF trial of mesenchymal precursor cells in chronic heart failure. Circ Res. 2019;125(3):265–81. https://doi.org/10.1161/CIRCRESAHA.119.314951. Epub 2019 Jul 18. Review. Erratum in: Circ Res. 2019 Aug 16;125(5):e28.

Part II
Heart Failure (Europe)

Chapter 6
Allogeneic Stem Cell Therapy in Heart Failure

Brief Summary:
The present aim is to perform a clinical double-blind, placebo-controlled Cardiology Stem Cell Center – Adipose Stem Cells (CSCC_ASC) study in heart failure patients to investigate the regenerative capacity of the CSCC_ASC treatment.

Condition or disease	Intervention/treatment	Phase
Heart failure	Biological: Cardiology stem cell center adipose stem cell (CSCC_ASC) biological: Placebo	Phase 2

Detailed Description:
The primary objective is to investigate the regenerative capacity of direct intra-myocardial injection of 100 mio. Allogeneic CSCC_ASCs in patients with reduced left ventricular ejection fraction (EF) ($\leq 45\%$) and heart failure in a double-blind placebo-controlled design.

A total of 81 patients with will be enrolled in the study and treated in a 2:1 randomization with either CSCC_ASC or placebo (saline).

The primary endpoint is change in the left ventricle end-systolic volume (LVESV) at 6 months follow-up.

Cardiac Catheterization Laboratory 2014, The Heart Centre, University Hospital, Rigshospitalet, Copenhagen Ø, Denmark
Cardiac Catheterization Laboratory 2014, The Heart Centre, University Hospital, Rigshospitalet, Copenhagen Ø, Denmark
ClinicalTrials.gov Identifier: NCT03092284.

Study Type:	Interventional (Clinical Trial)
Actual enrollment:	81 participants
Allocation:	Randomized
Intervention model:	Parallel assignment
Masking:	Double (participant, investigator)
Primary purpose:	Treatment
Official title:	Allogeneic adipose tissue-derived stromal/stem cell therapy in patients with ischemic heart disease and heart failure: A phase II Danish Multicenter Study
Actual study start date:	September 2015
Estimated primary completion date:	August 2020
Estimated study completion date:	February 2021

Arm	Intervention/treatment
Active comparator: Cardiology Stem Cell Center Adipose Stem Cell (CSCC_ASC) Allogeneic adipose-derived stromal cells	Biological: Cardiology stem cell center adipose stem cell (CSCC_ASC) Direct intramyocardial injection of CSCC_ASC
Placebo comparator: Placebo Saline	Biological: Placebo Saline

Primary Outcome Measures:
1. Change in left ventricle end-systolic volume (LVESV) from baseline to 6 months follow-up measured by echocardiography and computerized tomography [time frame: 6 months].

   Change in left ventricle end-systolic volume (LVESV) from baseline to 6 months follow-up measured by echocardiography and computerized tomography.

Secondary Outcome Measures:
1. Incidence of treatment-emergent adverse events [time frame: 12 months].
   Safety is evaluated by the incidence and severity of serious adverse events and suspected unrelated serious adverse events at 12 months follow-up.
2. Efficacy left Ventricle [Time Frame: 6 months]
   Change in left ventricle ejection fraction (EF) measured by echocardiography and computerized tomography.
3. Efficacy clinical function assessed by change in Kansas City Cardiomyopathy Questionnaire (KCCQ) from baseline to 6 months follow-up [Time Frame: 6 months].
   Change in Kansas City Cardiomyopathy Questionnaire (KCCQ) from baseline to 6 months follow-up.

4. Efficacy clinical function assessed by change in Seattle Angina Questionnaire and 6 minute walking test from baseline to 6 months follow-up [Time Frame: 6 months].

   Change in Seattle Angina Questionnaire and 6-minute walking test from baseline to 6 months follow-up.

5. Efficacy clinical function assessed by change in 6 minute walking test from baseline to 6 months follow-up [Time Frame: 6 months].

   Change in 6-minute walking test from baseline to 6 months follow-up.

6. Efficacy clinical function assessed by change in Kansas City Cardiomyopathy Questionnaire (KCCQ) from baseline to 12 months follow-up [Time Frame: 12 months].

   Change in Kansas City Cardiomyopathy Questionnaire (KCCQ) from baseline to 12 months follow-up.

7. Efficacy clinical function assessed by change in Seattle Angina Questionnaire and 6 minute walking test from baseline to 12 months follow-up [Time Frame: 12 months].

   Change in Seattle Angina Questionnaire and 6-minute walking test from baseline to 12 months follow-up.

8. Efficacy clinical function assessed by change in 6 minutr walking test from baseline to 12 months follow-up [Time Frame: 12 months].

   Change in 6-minute walking test from baseline to 12 months follow-up.

Ages eligible for study:	30 years to 80 years (adult, older adult)
Sexes eligible for study:	All
Accepts healthy volunteers:	No

Inclusion Criteria:
1. 30 to 80 years of age
2. Signed informed consent.
3. Chronic stable ischemic heart disease.
4. Symptomatic heart failure – New York Heart Association (NYHA) class II–III.
5. EF ≤ 45%.
6. Plasma NT-pro-BNP > 300 pg/mL (>35 pmol/L) in sinus rhythm and plasma NT-pro-BNP > 422 pg/mL (>450 pmol/L) in patients with atrial fibrillation.
7. Maximal tolerable heart failure medication.
8. Medication unchanged 2 months prior to inclusion.
9. No option for percutaneous coronary intervention (PCI) or coronary artery bypass graft (CABG).
10. Patients who have had PCI or CABG within 6 months of inclusion must have a new angiography less than 1 month before inclusion or at least 4 months after the intervention to rule out early restenosis.
11. Patients cannot be included until 3 months after implantation of a cardiac resynchronization therapy device.

Exclusion Criteria:
 1. Heart failure (NYHA I or IV).
 2. Acute coronary syndrome with elevation of creatine kinase (CK) isoenzyme MB (CKMB) or troponins, stroke, or transitory cerebral ischemia within 6 weeks of inclusion.
 3. Other revascularization treatment within 4 months of treatment.
 4. If clinically indicated, the patient should have a coronary angiography before inclusion.
 5. Moderate to severe aortic stenosis (valve area < 1.3 mm^2) or valvular disease with option for surgery.
 6. Diminished functional capacity for other reasons such as obstructive pulmonary disease (COPD) with forced expiratory volume (FEV) < 1 L/min, moderate to severe claudication, or morbid obesity.
 7. Clinical significant anemia (hemoglobin <6 mmol/L), leukopenia (leucocytes <2109/L), leukocytosis (leukocytes $>14,109$/L), or thrombocytopenia (thrombocytes $<50,109$/L).
 8. Anticoagulation treatment that cannot be paused during cell injections.
 9. Patients with reduced immune response.
10. History with malignant disease within 5 years of inclusion or suspected malignancy – except treated skin cancer other than melanoma.
11. Pregnant women.
12. Other experimental treatment within 4 weeks of baseline tests.
13. Participation in another intervention trial.

Chapter 7
Stem Cell Therapy in Advanced Heart Failure

Brief Summary:
Stem cell therapy may be a choice therapy for advanced heart failure patients refractory to medical therapy, internal cardiovertor with a defibrillator (ICD) recipients, and a previous history of myocardial infarction and coronary artery revascularization. These patients, without indication to receive a cardiac resynchronization therapy (CRT), may have a worsening of heart failure and symptoms of coronary artery disease. In this study, we have evaluated in 30 consecutive selected patients, the amelioration in failing heart NYHA class, hospitalization rate, echocardiographic left ventricle functionality, and the associated reduction in angina after a treatment with intra-thoracic stem cell infusion.

Condition or disease	Intervention/treatment	Phase
Congestive heart failure	Drug: Stem cell infusion	Phase 4

Study Type:	Interventional (Clinical Trial)
Estimated Enrollment:	30 participants
Allocation:	N/A
Intervention Model:	Single- Group Assignment
Masking:	None (Open Label)
Primary Purpose:	Diagnostic

Celestino Sardu, Naples, Italy
Raffaele Marfella, Naples, Italy
ClinicalTrials.gov Identifier: NCT02871466

© The Author(s), under exclusive license to Springer Nature
Switzerland AG 2022
J. N. Weiss, *Stem Cell Surgery Trials in Heart Failure and Diabetes*,
https://doi.org/10.1007/978-3-030-78010-4_7

Official Title:	Intra Cardiac Autologous Stem Cell Infusion May Improve Clinical Outcomes in Failing Heart Subjects Refractory to Maximal Drug Therapy and Internal Cardioverter with a Defibrillator (ICD) Recipients
Study Start Date:	January 2010
Estimated Primary Completion Date:	January 2022
Estimated Study Completion Date:	July 2022

Arm	Intervention/treatment
Experimental: Stem cell infusion	Drug: Stem cell infusion

Primary Outcome Measures:
1. NYHA class improvement [time frame: 12 months]
2. Reduction in hospitalization rate [time frame: 12 months]
3. Reduction in angina symptoms [time frame: 12 months]

Eligibility Criteria:

Ages Eligible for Study:	18 Years to 75 Years (Adult, Older Adult)
Sexes Eligible for Study:	All
Accepts Healthy Volunteers:	No

Inclusion Criteria:
- Heart failure post-coronary artery disease, refractory to maximal medical therapy, previous coronary artery surgical revascularization

Exclusion Criteria:
- Inflammatory chronic diseases, neoplastic diseases

Chapter 8
Stem Cell Therapy in Ischemic Non-treatable Cardiac Disease

Brief Summary:
The aim of the SCIENCE study, an international multicenter, double-blind, placebo-controlled study, is to investigate the efficacy of direct intra-myocardial injection of 100 mio. allogeneic Cardiology Stem Cell Center adipose-derived stem cells (CSCC_ASCs) in patients with reduced left ventricular ejection fraction (EF) (≤45%) and heart failure.

Condition or disease	Intervention/treatment	Phase
Heart failure	Biological: CSCC_ASC	Phase 2

Detailed Description:
The aim of the SCIENCE study, an international multicenter, double-blind, placebo-controlled study, is to investigate the efficacy of direct intra-myocardial injection of 100 mio. allogeneic Cardiology Stem Cell Center adipose-derived stem cells (CSCC_ASCs) in patients with reduced left ventricular EF (≤45%) and heart failure.

The primary objective is to investigate the regenerative capacity of direct intra-myocardial injection of 100 mio. allogeneic CSCC_ASCs in patients with reduced left ventricular EF (≤45%) and heart failure in a double-blind placebo-controlled design.

A total of 138 patients will be enrolled in the study and treated in a 2:1 randomization with either CSCC_ASC or placebo (saline).

2014 Department of Cardiology, The Heart Centre, University Hospital Rigshospitalet, Copenhagen, Denmark
ClinicalTrials.gov Identifier: NCT02673164

Study Type:	Interventional (Clinical Trial)
Actual Enrollment:	133 participants
Allocation:	Randomized
Intervention Model:	Parallel Assignment
Masking:	Quadruple (Participant, Care Provider, Investigator, Outcomes Assessor)
Primary Purpose:	Treatment
Official Title:	Stem Cell Therapy in Ischemic Non-treatable Cardiac Disease – SCIENCE A European Multicentre Trial
Actual Study Start Date:	January 2017
Estimated Primary Completion Date:	July 2020
Estimated Study Completion Date:	December 2020

Arm	Intervention/treatment
Active comparator: CSCC_ASC Cardiology Stem Cell Center adipose-derived stem cells (CSCC_ASC)	Biological: CSCC_ASC The stem cells will be injected directly into the myocardium using the NOGA XP system (BDS, US) Other name: Cardiology Stem Cell Center adipose-derived stem cells
Placebo comparator: Placebo Saline	Biological: CSCC_ASC The stem cells will be injected directly into the myocardium using the NOGA XP system (BDS, US) Other name: Cardiology Stem Cell Center adipose-derived stem cells

Outcome Measures

Primary Outcome Measures:

1. Left ventricle end-systolic volume (LVESV) [time frame: 6 months]

The primary endpoint is change in left ventricle end-systolic volume (LVESV) from baseline to 6 months follow-up measured by ECHO, MR, and CT between CSCC_ASC and placebo-treated.

Secondary Outcome Measures:

1. Safety – Serious adverse events [time frame: 6 months]

Incidence and severity of serious adverse events and suspected unrelated serious adverse events at 12 months follow-up.

Eligibility Criteria

Ages Eligible for Study:	30 Years to 80 Years (Adult, Older Adult)
Sexes Eligible for Study:	All
Accepts Healthy Volunteers:	No

Inclusion Criteria:
1. 30–80 years of age
2. Signed informed consent
3. Chronic stable ischemic heart disease
4. Symptomatic heart failure New York Heart Association (NYHA) class II–III
5. EF ≤45% on echocardiography, computerized tomography (CT), or magnetic resonance imaging (MRI) scan
6. Plasma NT-pro-BNP >300 pg/mL (>35 pmol/L)
7. Maximal tolerable heart failure medication
8. Heart failure medication unchanged 2 months prior to inclusion. Changes in diuretics accepted
9. No option for percutaneous coronary intervention (PCI) or coronary artery bypass graft (CABG)
10. Patients who have had PCI or CABG within 6 months of inclusion must have a new coronary angiography less than 1 month before inclusion or at least 4 months after the intervention to rule out early restenosis
11. Patients cannot be included until 3 months after implantation of a cardiac resynchronization therapy device (CRTD) and until 1 month after an ICD unit

Exclusion Criteria:
1. Heart failure (NYHA class I or IV)
2. Acute coronary syndrome with acute reversible elevation of CKMB or troponins, stroke, or transitory cerebral ischemia within 6 weeks of inclusion. Constant elevated troponin due to renal failure, heart failure etc. does not exclude the patient from the study
3. Other revascularization treatment within 4 months of treatment
4. If clinically indicated, the patient should have a coronary angiography before inclusion
5. Moderate to severe aortic stenosis (valve area <1.3 cm^2) or valvular disease with option for surgery or interventional therapy
6. Aortic valve replacement with an artificial heart valve. However, a trans-septal treatment approach can be considered in these patients
7. If the patient is expected to be candidate for MitraClip therapy of mitral regurgitation in the 12 months follow-up period

8. Diminished functional capacity for other reasons such as obstructive pulmonary disease (COPD) with forced expiratory volume (FEV) <1 L/min, moderate to severe claudication, or morbid obesity
9. Clinically significant anemia (hemoglobin <6 mmol/L), leukopenia (leukocytes <2 109/L), leukocytosis (leukocytes >14 109/L), or thrombocytopenia (thrombocytes <50 109/L)
10. Reduced kidney function (estimated glomerular filtration rate (eGFR) <30 mL/min)
11. Left ventricular thrombus
12. Anticoagulation treatment that cannot be paused during cell injections. Patients can continue with platelet inhibitor treatment
13. Patients with reduced immune response or known anti-HLA (human leukocyte antigen) antibodies
14. History with malignant disease within 5 years of inclusion or suspected malignity – except treated skin cancer other than melanoma
15. Pregnant women
16. Other experimental treatment within 4 weeks of baseline tests
17. Participation in another intervention trial
18. Life expectancy less than 1 year
19. Known hypersensitivity to dimethyl sulfoxide (DMSO), penicillin, and streptomycin

Chapter 9
Infusion Intracoronary of Mononuclear Autologous Adult Non Expanded Stem Cells of Bone Marrow on Functional Recovery in Patients with Idiopathic Dilated Cardiomyopathy and Heart Failure

Brief Summary:

Clinical trial, phase IIb, double-blind, randomized, controlled study with placebo. There is sufficient preliminary evidence to consider intracoronary injection of bone marrow progenitor cells as a viable, safe, and beneficial treatment in patients with dilated cardiomyopathy, although the biological mechanism of action of bone marrow cells in the myocardium is not known. In this project, we propose to investigate comparatively and from a biological and clinical point of view the applicability of regenerative therapy with autologous bone marrow cells in patients with dilated cardiomyopathy.

Condition or disease	Intervention/treatment	Phase
Idiopathic dilated cardiomyopathy	Drug: Infusion of autologous mononuclear bone marrow cell drug: placebo infusion	Phase 2

Detailed Description:

The study population includes male and female patients with idiopathic dilated cardiomyopathy.

A total of 51 patients diagnosed with this disease are included. After inclusion, we will proceed to the random allocation to study group or control group in a 2:1 ratio, 34 patients in the treatment group and 17 in the control group.

The total duration is expected to be 48 months: The inclusion period is 24 months, and each patient assigned to the experimental group will be followed for 24 months, whereas those randomized to the control group will have a follow-up

Hospital Universitario Puerta del Mar, Cádiz, Spain
Hospital Universitario Reina Sofía, Córdoba, Spain
Hospital Universitario Virgen de las Nieves, Granada, Spain
(and 6 more...)
ClinicalTrials.gov Identifier: NCT02033278

© The Author(s), under exclusive license to Springer Nature Switzerland AG 2022
J. N. Weiss, *Stem Cell Surgery Trials in Heart Failure and Diabetes*,
https://doi.org/10.1007/978-3-030-78010-4_9

43

period of 12 months. Upon completion, the patients will be followed up in routine clinical practice.

This is a double-blind study in which all patients will perform the bone marrow harvesting.

All patients will receive the best medical treatment individualized (ACEIs or angiotensin II receptor blocker, beta-blockers, diuretics, and eplerenone) for at least 6 months prior to their participation in the clinical trial, so that the situation is stable and pharmacological basal condition is the same for everyone.

The bone marrow cells of patients assigned to the placebo group will be cryopreserved, and once the trial is completed, the blinded group will be opened, and all patients who had been randomized to the control group may be included for compassionate use with their own, previously frozen, mononuclear bone marrow cells.

The patients who are randomized to the experimental group will be treated with the conventional treatment + infusion of autologous mononuclear bone marrow cells that are not expanded, whereas the patients who are randomized to the control group will be treated with the conventional treatment + infusion of placebo.

Primary Objective:
The primary objective is to assess the efficacy of intracoronary injection of autologous bone marrow stem cells to improve ventricular function in patients with idiopathic dilated cardiomyopathy who receive conventional medical treatment compared with the control group receiving an infusion of placebo and conventional medical treatment. The improvement in ventricular function is assessed by changes in angiographically determined ejection fraction.

Secondary Objectives:
The secondary objective is to analyze the predictors of good clinical response, functional and biological treatment, with unexpanded adult stem cells autologous mononuclear bone marrow cells.

The following parameters were evaluated: functional class (NYHA), natriuretic peptide B, stress test (exercise time), echocardiographic parameters of ventricular function (e.g., LVEF (%), TDV (mL), TSV (mL), and TAPSE (ms), and biological parameters of cellular functionality (e.g., CD133+, CD34+, CD34 +/CD177+, and CD34+/CD38− (in %).

The other objective is to determine, in the light of the obtained results, the cell therapy suitable for the application protocol for the treatment of dilated cardiomyopathy.

Study Type:	Interventional (Clinical Trial)
Estimated Enrollment:	27 participants
Allocation:	Randomized
Intervention Model:	Parallel Assignment
Masking:	Quadruple (Participant, Care Provider, Investigator, Outcomes Assessor)
Primary Purpose:	Treatment

Official Title:	Multicenter Phase IIb Clinical, Double-Blind, Randomized, Placebo-Controlled Trial to Assess the Efficacy of Intracoronary Infusion of Autologous Adult Stem Cell Mononuclear Marrow Unexpanded on Functional Recovery in Patients with Idiopathic Dilated Cardiomyopathy and Heart Failure.
Actual Study Start Date:	January 6, 2014
Estimated Primary Completion Date:	April 2020
Estimated Study Completion Date:	April 2020

Arm	Intervention/treatment
Experimental: Infusion of autologous mononuclear bone marrow cells Infusion of autologous mononuclear bone marrow cells plus conventional medical treatment (as indicated by the clinician)	Drug: Infusion of autologous mononuclear bone marrow cells. Infusion of autologous mononuclear bone marrow cells plus conventional medical treatment (as indicated by the clinician)
Placebo comparator: Placebo infusion Placebo infusion plus conventional medical treatment (as indicated by the clinician)	Drug: Placebo infusion Placebo infusion plus conventional medical treatment (as indicated by the clinician)

Outcome Measures
Primary Outcome Measures:
1. Changes in ventricular function measured angiographically [time frame: 24 months]

Secondary Outcome Measures:
1. Degree of clinical improvement based on the absence of major cardiac events (MACE) during follow-up [time frame: 24 months]
2. Clinical and analytical progress (NYHA grade and BNP) [time frame: 24 months]
3. Time of evolution since diagnosis of idiopathic dilated cardiomyopathy prior to study entry [time frame: 24 months]
4. Functional recovery as measured with ergometry [time frame: 24 months]
5. Echocardiography and electrocardiography variables [time frame: 24 months]

Eligibility Criteria:

Ages Eligible for Study:	18 Years to 70 Years (Adult, Older Adult)
Sexes Eligible for Study:	All
Accepts Healthy Volunteers:	No

Inclusion Criteria:
1. Patients of both sexes and ages between 18 and 70 years
2. Patients diagnosed with dilated cardiomyopathy via echocardiography

3. Minimum evolution since diagnosis of 6 months
4. Absence of coronary injury tested with multislice CT and/or hemodynamic study performed after study entry or within the previous 36 months (or before in specific low-risk clinical profiles) if no angina symptomatology is present
5. Patients receiving optimized medical therapy for at least 6 months prior to enrollment (individually adjusted according to functional status)
6. Ejection fraction of the left ventricle <40% or ejection fraction of the left ventricle 40–50% if the left ventricular telediastolic volume is >110 mL/m^2
7. Presence of sinus rhythm
8. Written informed consent for participation in the trial
9. Normal laboratory parameters, defined by the following: leukocytes ≥3000, neutrophils ≥1500, platelets ≥100,000, aspartate aminotransferase/alanine aminotransferase ≤2.5 standard range institution, creatinine ≤2.5 mg/dL, hemoglobin >9 g/dL
10. Women of childbearing potential must have negative results on a pregnancy test and agree to use medically approved methods of contraception throughout the follow-up.

Exclusion Criteria:
1. Secondary dilated cardiomyopathy.
2. Recent history of myocarditis (<6 months prior to study entry)
3. Patients amenable to receive cardiac resynchronization therapy
4. Patients in active waiting list for heart transplantation
5. Coexistence of other serious systemic diseases
6. Coexistence of any type of blood diseases
7. Pregnant or breastfeeding women or women of childbearing potential not committing to use effective contraception
8. Patients who are currently participating or have completed their participation in a clinical trial within the last 3 months and patients who have participated in any advanced therapy clinical trial any time previously
9. Patients with malignant or pre-malignant tumors
10. Positive serology for hepatitis B virus, hepatitis C virus, or human immunodeficiency virus
11. Use of any protocol-prohibited medication. A wash-out period of 2 months can be considered for inclusion in the trial.

Publications:
Romero M, Suárez-de-Lezo J, Herrera C, Pan M, López-Aguilera J, Suárez-de-Lezo J Jr, Baeza-Garzón F, Hidalgo-Lesmes FJ, Fernández-López O, Martínez-Atienza J, Cebrián E, Martín-Palanco V, Jiménez-Moreno R, Gutiérrez-Fernández R, Nogueras S, Carmona MD, Ojeda S, Cuende N, Mata R. Randomised, double-blind, placebo-controlled clinical trial for evaluating the efficacy of intracoronary injection of autologous bone marrow mononuclear cells in the improvement of the ventricular function in patients with idiopathic dilated myocardiopathy: a study protocol. BMC Cardiovasc Disord. 2019;19(1):203. https://doi.org/10.1186/s12872-019-1182-4.

Chapter 10
Cell Therapy in HFpEF

Brief Summary:
The primary objective of the study is to investigate the safety and efficacy of transendocardial CD34+ cell therapy in patients with HFpEF by evaluating changes in myocardial structure and function, myocardial perfusion and electrical activity, biomarkers of neurohormonal activation, patient exercise capacity, and clinical outcome.

The secondary objective of the study is to better define the pathophysiological background of HFpEF by using multimodality imaging platform, which will include data from electro-anatomical mapping, cardiac MRI, 2D and 3D echocardiography, high-resolution electrocardiography, and cardiac catheterization.

Condition or disease	Intervention/treatment	Phase
Heart failure with normal ejection fraction	Biological: Cell Therapy Biological: Control Procedure	Phase 2

Study Type:	Interventional (Clinical Trial)
Estimated Enrollment:	30 participants
Allocation:	Randomized
Intervention Model:	Parallel Assignment
Masking:	Quadruple (Participant, Care Provider, Investigator, Outcomes Assessor)
Primary Purpose:	Treatment
Official Title:	A Pilot Trial of Cell Therapy in Heart Failure with Preserved Ejection Fraction
Actual Study Start Date:	January 2016
Estimated Primary Completion Date:	January 2020
Estimated Study Completion Date:	March 2020

University Medical Center Ljubljana, Ljubljana, Slovenia
ClinicalTrials.gov Identifier: NCT02923609

© The Author(s), under exclusive license to Springer Nature
Switzerland AG 2022
J. N. Weiss, *Stem Cell Surgery Trials in Heart Failure and Diabetes*,
https://doi.org/10.1007/978-3-030-78010-4_10

Arm	Intervention/treatment
Active comparator: SC Group After enrollment, all patients will receive 5-day stem cell mobilization with G-CSF. Thereafter, the patients will be randomly allocated to either active (SC Group) or control group (Controls) in a 2:1 ratio. The patients in the SC Group will undergo apheresis; CD34+ cells will be collected via immunomagnetic selection and delivered transendocardially to the target areas defined by electro-anatomical mapping	Biological: Cell therapy electro-anatomical mapping will be performed using the Biosense NOGA system (Biosense Webster, Diamond Bar, California). Local diastolic function will be assessed by a novel algorithm that allows for the measurement of local ventricular relaxation times at each of the sampling points. Target areas for cell delivery will be defined as the myocardial segments with the evidence of local diastolic dysfunction and myocardial hibernation. Transendocardial delivery of cell suspension in the SC Group will be performed using MyoStar® (Biosense Webster) injection catheter. Each patient will receive 20 injections of 0.3 mL of stem cell suspension
Placebo comparator: Control In the Control group, no apheresis or immunomagnetic selection of CD34+ cells will be performed; the patients will receive transendocardial injections of placebo using the same electro-anatomical mapping protocol as that in patients from the SC Group	Biological: Control procedure In the Control Group, we will perform transendocardial injections using the same protocol as that in the SC Group; stem cell suspension will be substituted with placebo (0.9% saline)

Outcome Measures
Primary Outcome Measures:
1. Change in diastolic function (E/e′) assessed by cMRI [time frame: baseline and 1 year]

Secondary Outcome Measures:
1. Change in exercise capacity [time frame: baseline and 1 year]
2. Change in NT-proBNP levels [time frame: baseline and 1 year]

Eligibility Criteria:

Ages Eligible for Study:	18 Years to 70 Years (Adult, Older Adult)
Sexes Eligible for Study:	All
Accepts Healthy Volunteers:	No

Inclusion Criteria:
- Preserved left ventricular systolic function on echocardiography (LVEF >50%)
- Evidence of diastolic dysfunction by echocardiography (E/e′ >15)
- Symptoms of heart failure (NYHA functional class II or III)
- NT-proBNP levels > 300 pg/mL

Exclusion Criteria:
- Acute multi-organ failure
- History of any malignant disease within 5 years
- Diminished functional capacity due to non-cardiac co-morbidities (COPD, PAOD, morbid obesity)
- Pregnancy

Chapter 11
Compare the Effects of Single Versus Repeated Intracoronary Application of Autologous Bone Marrow-Derived Mononuclear Cells on Mortality in Patients with Chronic Post-infarction Heart Failure

Brief Summary:

Single or repeated application of autologous bone marrow-derived stem cells to treat chronic post-infarction heart failure

Condition or disease	Intervention/treatment	Phase
Heart failure	Biological: intracoronary infusion of autologous bone marrow-derived cells	Phase 2 Phase 3

Detailed Description:

Improve mortality and morbidity in patients with symptomatic chronic post-infarction heart failure under full-dose conventional medical and device treatment, including resynchronization therapy, via single versus repeated intracoronary infusion of autologous bone marrow-derived mononuclear cells.

Study Type:	Interventional (Clinical Trial)
Actual Enrollment:	81 participants
Allocation:	Randomized
Intervention Model:	Parallel Assignment
Masking:	None (Open Label)
Primary Purpose:	Treatment

Zentralklinik Bad Berka, Bad Berka, Germany
Goethe University Frankfurt, Frankfurt, Germany
Klinikum Fulda, Fulda, Germany
(and 3 more...)
ClinicalTrials.gov Identifier: NCT01693042

Official Title:	Randomized Controlled Trial to Compare the Effects of Single Versus Repeated Intracoronary Application of Autologous Bone Marrow-Derived Mononuclear Cells on Total and SHFM-Predicted Mortality in Patients with Chronic Post-Infarction Heart Failure
Study Start Date:	November 2013
Estimated Primary Completion Date:	November 2022
Estimated Study Completion Date:	January 2025

Arms and Interventions

Arm	Intervention/treatment
Active comparator: Single intracoronary cell application Single intracoronary application of autologous bone marrow-derived mononuclear cells	Biological: Intracoronary infusion of autologous bone marrow-derived cells Intracoronary infusion into open vessel/bypass supplying previous (> 3 months) infarct area Other name: t2c001
Active comparator: Repeated (2 times) intracoronary cell application 2 times (interval 4 months) intracoronary application of autologous bone marrow-derived mononuclear cells	Biological: Intracoronary infusion of autologous bone marrow-derived cells Intracoronary infusion into open vessel/bypass supplying previous (>3 months) infarct area Other name: t2c001

Primary Outcome Measures:
1. Mortality at 2 years after inclusion into the study [time frame: 2 years]

   2-year observed mortality is significantly lower in patients receiving 2 repeated intracoronary applications of autologous bone marrow-derived cells (t2c001) compared with patients receiving 1 intracoronary application of autologous bone marrow-derived cells (t2c001)

Secondary Outcome Measures:
1. Morbidity at 2 and 5 years after inclusion into the study [time frame: 2 years and 5 years]

Efficacy Endpoints:
Comparison between the 2 treatment groups at 2-year and 5-year follow-ups:

- Cardiac mortality, cardiovascular mortality
- Rehospitalization for heart failure
- Ischemic cardiac events (STEMI, NSTEMI, ACS)
- Coronary revascularizations (PCI/CABG)
- Heart transplantation, assist device implantation
- New resynchronization therapy, ICD implantation

- NYHA-Status, NT-proBNP serum levels
- Minnesota Living with Heart Failure Questionnaire

Safety Endpoints:
Bleeding events, all in-hospital events (during hospitalization for BMC therapy), life-threatening arrhythmias, new malignancies

Eligibility Criteria:

Ages Eligible for Study:	18 Years to 80 Years (Adult, Older Adult)
Sexes Eligible for Study:	All
Accepts Healthy Volunteers:	No

Inclusion Criteria:
- Previous myocardial infarction at least 3 months ago, open infarct vessel or bypass
- Left ventricular ejection fraction (LVEF) ≤45% on echocardiography
- Stable chronic heart failure NYHA classes II to III under constant (4 weeks) evidence-based optimal medical treatment
- Aged 18–80 years
- Written informed consent
- Women of childbearing potential: negative pregnancy test; effective contraception for the first 8 months in the trial

Exclusion Criteria:
- Non-ischemic cardiomyopathy
- Necessity for revascularization in other vessel than the infarct vessel at the time of study therapy
- Hemodynamic relevant severe valvular disease with indication for operative/interventional revision
- Heart failure with preserved ejection fraction (diastolic heart failure), LVEF >45%
- Unstable angina
- Severe peripheral artery occlusive disease (≥Fontaine stadium III)
- Active infection (C-reactive protein >10 mg/dL), chronic active hepatitis; any chronic inflammatory disease, HIV infection
- Neoplastic disease without documented remission in the last 5 years
- Stroke ≤3 months
- Impaired renal function (serum creatinine >2.5 mg/dL) at the time of study inclusion
- Relevant liver disease (GOT >2× upper normal limit, spontaneous INR >1.5)
- Diseases of hematopoietic system, anemia (hemoglobin <8.5 mg/dL), thrombocytopenia <100,000/μL)
- Splenomegaly
- Allergy or intolerance of clopidogrel, prasugrel, ticagrelor, heparin, bivalirudin
- History of bleeding disorder
- Gastrointestinal bleeding ≤3 months

- Major surgery or trauma ≤3 months
- Uncontrolled hypertension
- Pregnancy, lactation period
- Mental retardation
- Previous cardiac cell therapy within the last 12 months
- Participation in another clinical trial ≤30 days

Chapter 12
Impact of Intracoronary Injection of Autologous BMMC for LV Contractility and Remodeling in Patients with STEMI

Brief Summary:

This is a multicenter, randomized, open-label, controlled, parallel-group phase III study. Its aim is to demonstrate that a triple intracoronary infusion of autologous bone marrow-derived mononuclear cells in addition to state-of-the-art treatment is safe and reduces all-cause mortality in patients with reduced left ventricular ejection fraction ($\leq 45\%$) after successful reperfusion for acute myocardial infarction when compared with a control group comprising of patients undergoing the best medical care.

Condition or disease	Intervention/treatment	Phase
Heart failure myocardial infarction	Procedure: Intracoronary infusion of BM-MC	Phase 3

Detailed Description:

The study is divided into three parts:

- Screening phase: Patients will be recruited at the investigational clinical centers. Alternatively, patients who had primary PCI performed at institutions different from the investigational sites can also be enrolled. Interested patients may be referred for screening to any of the participating study sites after acute reperfusion therapy. Informed consent and assessment of eligibility of patients with respect to inclusion and exclusion criteria will be done at the investigational site. If all other eligibility criteria are met, echocardiography will be performed 3–6 days after the acute PCI, and ejection fraction will be quantified by a central Echo Core Lab after web-based transmission. CT examination will be performed

Polsko-Amerykańskie Kliniki Serca, Ustroń, Poland
ClinicalTrials.gov Identifier: NCT02323620

1 month after acute PCI in all screened patients with LVEF ≤45%. If LVEF will not improve by ≥5% in the CT, the patient may qualify to the study.

- Treatment phase: Bone marrow aspiration will be performed for the patients assigned to the treatment group (II). Bone marrow will be collected from the patient and MNC isolated using point-of-care system (Harvest) at a site. Intracoronary infusion of BM-MNCs will be performed up to 2 hours after isolation via radial approach. The same procedure will be performed 3 and 6 months after the first application.
- Follow-up phase: After hospital discharge, patients will be followed up via telephone 30 days and 3, 6, and 9 months after randomization and with a site visit with CT examination 12 months after randomization. Afterwards, telephone follow-up will be performed every 3 months. Once the required number of clinical events has been observed, all patients will attend a final study visit, but the minimum follow-up period for each patient is 2 years. Endpoints will be reported as occurring throughout the follow up.

Study Type:	Interventional (Clinical Trial)
Estimated Enrollment:	200 participants
Allocation:	Randomized
Intervention Model:	Parallel Assignment
Masking:	None (Open Label)
Primary Purpose:	Treatment
Official Title:	The Impact of Repeated Intracoronary Injection of Autologous Bone Marrow-Derived Mononuclear Cells for Left Ventricle Contractility and Remodeling in Patients with STEMI Prospective Randomized Study
Estimated Study Start Date:	March 2019
Estimated Primary Completion Date:	July 2022
Estimated Study Completion Date:	December 2022

Arms and Interventions

Arm	Intervention/treatment
No Intervention: Standard care Optimal standard care after myocardial infarction	
Experimental: Intracoronary infusion of BM-MC Bone marrow-derived progenitor autologous cell aspiration and intracoronary infusion of the cells.	Procedure: Intracoronary infusion of BM-MC Bone marrow-derived progenitor cells are obtained from 60-mL bone marrow aspirated from the iliac crest. Intracoronary infusion of the autologous cells is performed via conventional percutaneous intracoronary intervention techniques using an over-the-wire balloon technique

Outcome Measures
Primary Outcome Measures:
1. Left ventricle ejection fraction change evaluated by CT [time frame: 12 months]

Secondary Outcome Measures:
1. Change in left ventricle end-systolic volume (ESV) and end-diastolic volume (EDV) evaluated by CT [time frame: 12 months]
2. Time from randomization to cardiac death [time frame: 3 years]
3. Time from randomization to cardiovascular death or rehospitalization due to heart failure [time frame: 3 years]
4. Incidence and severity of adverse events [time frame: 3 years]

Eligibility Criteria

Ages Eligible for Study:	18 Years and older (Adult, Older Adult)
Sexes Eligible for Study:	All
Accepts Healthy Volunteers:	No

Inclusion Criteria:
1. Men and women of any ethnic origin aged $\geq$18 years.
2. Patients with acute ST-elevation myocardial infarction as defined by the universal definition of AMI
3. Successful acute reperfusion therapy (residual stenosis visually <50% and TIMI flow $\geq$2) within 24 hours of symptom onset or thrombolysis within 12 hours of symptom onset followed by successful percutaneous coronary intervention (PCI) within 24 hours after thrombolysis
4. Left ventricular ejection fraction $\leq$45% with significant regional wall motion abnormality assessed by quantitative echocardiography (central, independent core lab analysis) 3–6 days after reperfusion therapy
5. Open coronary artery suitable for cell infusion supplying the target area of abnormal wall motion
6. LVEF $\leq$5% with significant regional wall motion abnormality assessed by computed tomography (CT) 30 days after reperfusion therapy with no LVEF improvement $\geq$5%

Exclusion Criteria:
1. Participation in another clinical trial within 30 days prior to randomization
2. Previously received stem/progenitor cell therapy
3. Pregnant or nursing women
4. Mental condition rendering the patient unable to understand the nature, scope, and possible consequences of the study or to follow the protocol
5. Necessity to revascularize additional vessels outside the target coronary artery at the time of BM-MNC infusion (additional revascularizations after primary PCI and before BM-MNC cell infusion are allowed)
6. Cardiogenic shock requiring mechanical support

7. Platelet count <100,000/µL or hemoglobin <8.5 g/dL
8. Impaired renal function, i.e., serum creatinine >2.5 mg/dL
9. Persistent fever or diarrhea not responsive to treatment within 4 weeks prior to screening
10. Clinically significant bleeding within 3 months prior to screening
11. Uncontrolled hypertension (systolic >180 mmHg and diastolic >120 mmHg)
12. Life expectancy of less than 2 years from any non-cardiac cause or neoplastic disease

Chapter 13
Stem Cell Therapy in Non-ischemic, Non-treatable Dilated Cardiomyopathies II: A Pilot Study

Brief Summary:
The overall aim of the project is to test the feasibility and safety of allogeneic adipose-derived stromal cell (CSCC_ASC) investigational medicinal product to improve myocardial function in patients with non-ischemic dilated cardiomyopathies (NIDCM) and heart failure.

Condition or disease	Intervention/treatment	Phase
Non-ischemic dilated cardiomyopathy	Biological: Allogeneic adipose-derived stromal cells (CSCC_ASC) Other: Control group	Phase 1 Phase 2

Detailed Description:

Study design
- The primary objective of the study is to investigate safety and regenerative capacity of direct intra-myocardial injection of 100 million allogeneic CSCC_ASCs in NIDCM patients with reduced left ventricular EF ($\leq$40%) and heart failure.
- It is a proof-of-concept study enrolling a total of 30 NIDCM patients with heart failure who will be randomly allocated in a 2:1 ratio to either CSCC_ASC cell therapy (Stem Cell Group) or no cell therapy (Control Group). The treatment period is estimated to be 6 months (efficacy endpoint) with a 12-month follow-up period for safety endpoints.

Patient treatment and follow-up:
- The cell IMP will be delivered by a courier service from REGIONH to UKCL using validated portable dry liquid nitrogen shipping containers. It will then be stored in nitrogen vapor containers until treatment.

The Heart Centre, Rigshospitalet University Hospital Copenhagen, Copenhagen, Denmark
ClinicalTrials.gov Identifier: NCT03797092

- The preparation of the IMP will be performed as described in the treatment manual. The IMP will be thawed and prepared for injection immediately before treatment.
- A 3D map of the left ventricle will be created using the NOGA XP® system (Biological Delivery System, Cordis, Johnson & Johnson, USA). The delivery of the IMP (100 million ASCs) to the myocardium will be performed via 10–15 injections of 0.2 cc, as described in the treatment manual.
- The post cell therapy surveillance will include clinical and laboratory safety follow-up with monitoring of cardiac enzymes and hemodynamic and rhythm stability. The clinical follow-up will be obtained at pre-defined time points. The presence of allogeneic HLA anti-bodies will be monitored 3 and 6 months after treatment. Endpoints will be monitored continuously and reported as occurring throughout the 12-month follow-up period.

Primary and secondary endpoints:
- The primary endpoint is change in left ventricle end-systolic volume (LVESV) at 6 months follow-up measured via echocardiography.
- The secondary endpoints are safety evaluated through the development of allogeneic antibodies and laboratory safety measurements 1, 3, and 6 months after treatment and changes in left ventricular ejection fraction (LVEF), end-diastolic volume, and myocardial mass at 6 months follow-up.
- Additional secondary endpoints are changes in NYHA, Kansas City Cardiomyopathy Questionnaire, EQ-5D3L Questionnaire, 6-minute walking test, additional echocardiographic measures (global strain %), and NT-pro-BNP.
- Safety of allogeneic CSCC_ASCs with respect to incidence and severity of serious adverse events and suspected unrelated serious adverse events will be evaluated at 12 months follow-up.
- Outcome measures for safety endpoints will be collected, reported to the authorities, and monitored according to the legislation during the entire study period.
- Adverse event (AE) is defined as any untoward medical occurrence in a subject who was treated with an investigational product and does not necessarily have a causal relationship with the treatment. An AE can therefore be any unfavorable and unintended sign, symptom, or disease, whether or not related to the investigational product.

Serious adverse event (SAE) is defined as any untoward medical occurrence that:

1. Results in death
2. Is life-threatening
3. Requires inpatient hospitalization or prolongation of existing hospitalization
4. Results in persistent or significant disability/incapacity
5. Is a congenital anomaly/birth defect
6. Is medically important

Suspected unexpected serious adverse event reactions (SUSAR) is defined as an SAE occurring in a subject in an interventional study that is assessed as both causally related to the suspect product under clinical investigation and unexpected per the Investigator's Brochure (IB).

An independent Data Safety Monitoring Board (DSMB) will be established to evaluate the safety of the treatment. The DSMB will report directly to the director of the Project Management Board, which will take the necessary action upon the DSMB's recommendations.

Study medication: The investigational cell product, CSCC_ASC, will be produced in an approved GMP facility in Cardiology Stem Cell Center at Rigshospitalet University Hospital, Copenhagen, Denmark.

The production of the allogeneic CSCC_ASCs will follow the description in an approved Investigational Medicinal Product Dossier. The cell product will come from healthy donors.

The production unit will label the investigational medicinal product (IMP) in accordance with the legislation and keep the randomization code until finalization of the clinical trial. The final cell products will be stored in nitrogen vapor containers until clinical use.

Allogeneic MSCs and ASCs have been administered to more than 600 patients with heart disease. In the conducted clinical trials, there has not been any serious adverse event due to the treatment. A few patients have developed transient donor-specific HLA-antibodies in serum within the first months after treatment. However, none of the patients had any symptoms related to the presence of antibodies. Transient fever was registered in a few patients, but it could as well be due to the treatment procedure or the disease for treatment.

Based on the accumulated safety and efficacy evidence with the clinical use of allogeneic MSCs and ASCs in the conducted clinical trials and the safety data from the CSCC_ASC phase I trial and the two ongoing phase II trials, it is safe to conduct a pilot CSCC_ASC trial in patients with NIDCM and HF.

Echocardiography
- The echocardiography data will be recorded at pre-defined intervals according to the American Society of Echocardiography (ASE) and European Association of Cardiovascular Imaging (EACVI) recommendations. For each patient, at least five end-expiratory full cardiac cycles will be recorded for each protocol-specified view. All the acquired images will be de-identified and transferred to independent imaging core lab (Stanford Cardiovascular Institute Clinical Biomarker and Phenotype Core Lab). The recordings will be analyzed at the end of the study by an independent echocardiographer who will be blinded to the patient's treatment status and the timing of the recordings. All measurements will be performed according to the ASE/EACVI recommendations. All echocardiographic measurements will be averaged over 5 cardiac cycles. Left ventricular end-systolic dimension (LVESD) and end-diastolic dimension (LVEDD) will be measured in the parasternal long-axis view. Left ventricular end-systolic volume (LVESV), left ventricular end-diastolic volume (LVEDV), and LVEF will be estimated using Simpson's biplane method. Peak longitudinal strains will be computed automatically to generate regional data from each of the 17 segments and then averaged to calculate the global longitudinal strain.

Study Type:	Interventional (Clinical Trial)
Estimated Enrollment:	30 participants
Allocation:	Randomized
Intervention Model:	Parallel Assignment
Intervention Model Description:	Open randomized treatment group and control group clinical trial
Masking:	Single (Outcomes Assessor)
Masking Description:	The patients will be randomized to either IMP or control in a 2:1 randomization. The outcome ECHO investigations will be analyzed blinded by an independent core lab.
Primary Purpose:	Treatment
Official Title:	Stem Cell Therapy in Non-ischemic, Non-treatable Dilated Cardiomyopathies II: A Pilot Study
Actual Study Start Date:	October 1, 2019
Estimated Primary Completion Date:	September 1, 2020
Estimated Study Completion Date:	September 1, 2021

Arms and Interventions

Arm	Intervention/treatment
Active comparator: Active Allogeneic adipose-derived stromal cells (CSCC_ASC)	Biological: Allogeneic adipose-derived stromal cells (CSCC_ASC) Active group Other name: Investigational medicinal product Other: Control group No treatment
No Intervention: Control group No treatment	

Outcome Measures
Primary Outcome Measures:
1. Left ventricle end-systolic volume [time frame: 6 months after treatment]

   Measured via echocardiography

Secondary Outcome Measures:
1. Allogeneic antibodies [time frame: up to 12 months after treatment]

   Development of allogeneic antibodies and laboratory safety measurements 1, 3, and 6 months after treatment

2. Left ventricular ejection fraction [time frame: 6 months after treatment]

   Changes in LVEF

3. Myocardial mass of the left ventricle [time frame: 6 months after treatment]

   Change in echo measured global myocardial mass

4. NYHA [time frame: 6 months after treatment]

   Symptoms

5. Kansas City Cardiomyopathy Questionnaire [time frame: 6 months after treatment]

   Questionnaire

6. EQ-5D3L Questionnaire [time frame: 6 months after treatment]

   Questionnaire

7. 6-Minute walking test [time frame: 6 months]

   Test

Eligibility Criteria

Ages Eligible for Study:	30 Years to 80 Years (Adult, Older Adult)
Sexes Eligible for Study:	All
Accepts Healthy Volunteers:	No

Inclusion Criteria:
1. 30–80 years of age
2. Signed informed consent
3. Patients with non-ischemic dilated cardiomyopathy
4. NYHA ≥II despite the optimal heart failure treatment and the lack of other treatment options
5. Heart failure medication unchanged 2 months prior to inclusion/signature of informed consent. Changes in diuretics accepted
6. LVEF ≤405%
7. Plasma NT-pro-BNP >300 pg/mL (>35 pmol/L)
8. Patients cannot be included until 3 months after implantation of a cardiac resynchronization therapy device (CRTD) and until 1 month after an ICD unit

Exclusion Criteria:
1. Heart Failure NYHA I
2. Moderate to severe aortic stenosis (valve area <1.3 cm^2) or valvular disease with option for surgery or interventional therapy
3. Heart failure caused by cardiac valve disease or untreated hypertension
4. If the patient is expected to be candidate for MitraClip therapy of mitral regurgitation in the 12-month follow-up period
5. Cardiomyopathy with a reversible cause that has not been treated, e.g., thyroid disease, alcohol abuse, hypophosphatemia, hypocalcemia, cocaine abuse, selenium toxicity, and chronic uncontrolled tachycardia

6. Cardiomyopathy in association with a neuromuscular disorder, e.g., Duchenne progressive muscular dystrophy
7. Previous cardiac surgery
8. Diminished functional capacity for other reasons such as obstructive pulmonary disease (COPD) with forced expiratory volume (FEV) <1 L/min, moderate to severe claudication, or morbid obesity
9. Clinically significant anemia (hemoglobin <6 mmol/L), leukopenia (leukocytes <2 109/L), leukocytosis (leukocytes >14 109/L), or thrombocytopenia (thrombocytes <50 109/L)
10. Reduced kidney function (eGFR <30 mL/min)
11. Left ventricular thrombus
12. Anticoagulation treatment that cannot be paused during cell injections
13. Patients with reduced immune response
14. History with malignant disease within 5 years of inclusion or suspected malignancy, except treated skin cancer other than melanoma
15. Pregnant women
16. Woman of childbearing potential unless βHCG-negative; they should be on contraception during the trial
17. Other experimental treatment within 4 weeks of baseline tests
18. Participation in another intervention trial
19. Life expectancy of less than 1 year

Publications

Kastrup J, Haack-Sørensen M, Juhl M, Harary Søndergaard R, Follin B, Drozd Lund L, Mønsted Johansen E, Ali Qayyum A, Bruun Mathiasen A, Jørgensen E, Helqvist S, Jørgen Elberg J, Bruunsgaard H, Ekblond A. Cryopreserved Off-the-Shelf Allogeneic Adipose-Derived Stromal Cells for Therapy in Patients with Ischemic Heart Disease and Heart Failure-A Safety Study. Stem Cells Transl Med. 2017;6(11):1963–71. https://doi.org/10.1002/sctm.17-0040. Epub 2017 Sep 7.

Part III
Heart Failure (Iran)

Chapter 14
IMMNC-HF: IntraMyocardial Injection of Bone Marrow MonoNuclear Cells in Heart Failure (HF) Patients

Brief Summary:

This is a prospective, randomized, double-blind, controlled, parallel-group phase I and II study. Its aim is to assess whether the evaluation of the intramyocardial injection of autologous bone marrow-derived mononuclear stem cells is safe and effective in heart failure patients with reduced left ventricular ejection fraction (<=30%) when compared with a control group comprising of patients undergoing the best medical care.

Condition or disease	Intervention/treatment	Phase
Heart failure	Biological: Intramyocardial injection of stem cell Biological: Placebo	Phase 1 Phase 2

Study Design

Study Type:	Interventional (Clinical Trial)
Estimated Enrollment:	5 participants
Allocation:	Non-randomized
Intervention Model:	Parallel Assignment
Masking:	Double (Participant, Outcomes Assessor)
Primary Purpose:	Treatment
Official Title:	Evaluation of the Intramyocardial Injection of Autologous Bone Marrow-Derived Mononuclear Stem Cells in Heart Failure (HF) Patients

Stem Cell and Regenerative Medicine Institute (SCARM), Tabriz, Islamic Republic of Iran
ClinicalTrials.gov Identifier: NCT03227198

Estimated Study Start Date:	October 14, 2020
Estimated Primary Completion Date:	August 10, 2021
Estimated Study Completion Date:	October 12, 2021

Arms and Interventions

Arm	Intervention/treatment
Experimental: Intramyocardial injection of stem cell Intramyocardial injection of autologous bone marrow mononuclear cells in patients with heart failure	Biological: Intramyocardial injection of stem cell Autologous bone marrow-derived mononuclear cells are obtained from 100 to 150 mL of bone marrow aspirated under local anesthesia from the iliac crest. Intramyocardial injection of the cells is performed in patients with heart failure in open-heart surgery (CABG)
Placebo comparator: Placebo Placebo intramyocardial injection in patients with heart failure	Biological: Placebo Administration of placebo to patients with heart failure in open-heart surgery (CABG)

Outcome Measures
Primary Outcome Measures:
1. Death [time frame: 12 months]

   The rate of patient mortality after transplantation

2. Hospitalization [time frame: 12 months]

   The rate of hospitalization after transplantation

Secondary Outcome Measures:
1. Ejection fraction changes [time frame: 12 months]

   Elevation of ejection fraction in patients after transplantation

2. 6-Minute walk test (6MWT) [time frame: 12 months]

   Evaluation of the improvement of the 6MWT test after transplantation

3. Pro b-type natriuretic peptide (Pro-BNP) changes [time frame: 12 months]

   Evaluation of the reduction of Pro-BNP in patients after transplantation

4. NYHA functional class [time frame: 12 months]

   Evaluation of the improvement of NYHA functional class in patients

Eligibility Criteria

Ages Eligible for Study:	18 Years to 65 Years (Adult, Older Adult)
Sexes Eligible for Study:	All
Accepts Healthy Volunteers:	No

Inclusion Criteria:

1. Men and women of any ethnic origin $18 \le$ age ≤ 65 years
2. EF ≤ 40 (via echocardiography) and regional wall motion abnormality
3. Not responding to standard therapies
4. New York Heart Association (NYHA) class $\ge$III
5. Myocardial infarction due to coronary artery atherosclerotic disease
6. An area of regional dysfunction, i.e., hypokinetic, akinetic, or dyskinetic (echo-cardiography or MRI)
7. Normal liver and renal function
8. No or controlled diabetes
9. Able to give voluntary written consent and understand the study information provided

Exclusion Criteria:

1. Participation in another clinical trial within 30 days prior to randomization
2. Previously received stem/progenitor cell therapy
3. Pregnant women
4. Mental condition rendering the patient unable to understand the nature, scope, and possible consequences of the study or to follow the protocol
5. Cardiogenic shock requiring mechanical support
6. Congenital/valvular heart disease
7. Implantable cardioverter defibrillator (ICD) transplant
8. Platelet count <100,000/μL or hemoglobin <8.5 g/dL
9. Impaired renal function, i.e., creatinine >2.5 mg/dL
10. Fever or diarrhea within 4 weeks prior to screening
11. History of bleeding disorder within 3 months prior to screening
12. Uncontrolled hypertension (systolic >180 mmHg and diastolic >120 mmHg) or sustained ventricular arrhythmia
13. Life expectancy of less than 2 years from any non-cardiac cause or uncontrolled neoplastic disease

Chapter 15
ICMNC-HF: IntraCoronary Bone Marrow MonoNuclear Cells in Heart Failure (HF) Patients

Brief Summary:
This is a prospective, randomized, double-blind, controlled, parallel-group phase I and II study. Its aim is to assess whether a single intracoronary infusion of autologous bone marrow mononuclear cells is safe and effective in heart failure patients with reduced left ventricular ejection fraction (<=30%) when compared with a control group comprising of patients undergoing the best medical care.

Condition or disease	Intervention/treatment	Phase
Heart failure	Biological: Intracoronary injection of stem cell	Phase 1
	Biological: Placebo	Phase 2

Study Design

Study Type:	Interventional (Clinical Trial)
Estimated Enrollment:	5 Participants
Allocation:	Non-randomized
Intervention Model:	Parallel Assignment
Masking:	Double (Participant, Outcomes Assessor)
Primary Purpose:	Treatment
Official Title:	The Effect of Intracoronary Autologous Bone Marrow-Derived Mononuclear Stem Cell Transplantation in Heart Failure (HF) Patients
Estimated Study Start Date:	October 20, 2020

Stem Cell and Regenerative Medicine institute (SCARM), Tabriz, Islamic Republic of Iran
ClinicalTrials.gov Identifier: NCT03145402

Estimated Primary Completion Date:	August 28, 2021
Estimated Study Completion Date:	October 2, 2021

Arms and Interventions

Arm	Intervention/treatment
Experimental: Intracoronary injection of stem cell Autologous bone marrow-derived mononuclear cell injection to patients with heart failure	Biological: Intracoronary injection of stem cell Autologous bone marrow-derived mononuclear cells are obtained from 100 to 150 mL of bone marrow aspirated under local anesthesia from the iliac crest. Intracoronary infusion of the cells is performed in patients with heart failure using conventional percutaneous intracoronary intervention techniques
Placebo comparator: Placebo Placebo injection via coronary arteries in patients with heart failure	Biological: Placebo Injection of Placebo in patients with heart failure via coronary arteries

Outcome Measures
Primary Outcome Measures:
1. Death [time frame: 12 months]

   The rate of patient mortality after transplantation

2. Hospitalization [time frame: 12 months]

   The rate of hospitalization after transplantation

Secondary Outcome Measures:
1. Ejection fraction changes [time frame: 12 months]

   Elevation of ejection fraction in patients after transplantation

2. 6-Minute walk test (6MWT) [time frame: 12 months]

   Evaluation of the improvement of the 6MWT test after transplantation

3. Pro b-type natriuretic peptide (Pro-BNP) changes [time frame: 12 months]

   Evaluation of the reduction of Pro-BNP in patients after transplantation

4. NYHA functional class [time frame: 12 months]

   Evaluation of the improvement of the NYHA functional class in patients

Eligibility Criteria:

Ages Eligible for Study:	18 Years to 65 Years (Adult, Older Adult)
Sexes Eligible for Study:	All
Accepts Healthy Volunteers:	No

Inclusion Criteria:
1. Men and women of any ethnic origin $18 \leq$ age ≤ 65 years
2. EF ≤ 40 (via echocardiography) and regional wall motion abnormality
3. Not responding to standard therapies
4. New York Heart Association (NYHA) class $\geq$III
5. Myocardial infarction due to coronary artery atherosclerotic disease
6. An area of regional dysfunction, i.e., hypokinetic, akinetic, or dyskinetic (echo-cardiography or MRI)
7. Normal liver and renal function
8. No or controlled diabetes
9. Able to give voluntary written consent and understand the study information provided

Exclusion Criteria:
1. Participation in another clinical trial within 30 days prior to randomization
2. Previously received stem/progenitor cell therapy
3. Pregnant women
4. Mental condition rendering the patient unable to understand the nature, scope, and possible consequences of the study or to follow the protocol
5. Cardiogenic shock requiring mechanical support
6. Congenital/valvular heart disease
7. Implantable cardioverter defibrillator (ICD) transplant
8. Platelet count <100,000/µL or hemoglobin <8.5 g/dL
9. Impaired renal function, i.e., creatinine >2.5 mg/dL
10. Fever or diarrhea within 4 weeks prior to screening
11. History of bleeding disorder within 3 months prior to screening
12. Uncontrolled hypertension (systolic >180 mmHg and diastolic >120 mmHg) or sustained ventricular arrhythmia
13. Life expectancy of less than 2 years from any non-cardiac cause or uncontrolled neoplastic disease

Part IV
Heart Failure (China)

Chapter 16
Treating Heart Failure with hPSC-CMs

Brief Summary:

Heart failure has high morbidity and mortality rates because the heart is one of the least regenerative organs in the human body. Drug treatments for heart failure manage symptoms but do not restore the lost myocytes. Cellular replacement therapy is a potential approach for repairing damaged myocardial tissue and restoring cardiac function, which has become a new strategy for the treatment of heart failure. The purpose of this study is to assess the safety, feasibility, and efficacy of intramyocardial delivery of regenerated cardiomyocytes at the time of coronary artery bypass grafting in patients with chronic ischemic cardiomyopathy.

Condition or disease	Intervention/treatment	Phase
Heart failure	Biological: hPSC-CM therapy	Not applicable

Detailed Description:

Patients with heart failure will be treated with allogeneic human-induced pluripotent stem cell-derived cardiomyocytes (hPSC-CM) from healthy donors. The cells will be injected directly into the myocardium at the time of coronary artery bypass grafting. Patients will be assessed at 3, 6, and 12 months after cell transplantation for safety, feasibility, and efficacy.

Study Design

Study Type:	Interventional (Clinical Trial)
Estimated Enrollment:	5 Participants
Allocation:	N/A
Intervention Model:	Single-Group Assignment

HelpThera, Nanjing, Jiangsu, China
ClinicalTrials.gov Identifier: NCT03763136

J. N. Weiss, *Stem Cell Surgery Trials in Heart Failure and Diabetes*,
https://doi.org/10.1007/978-3-030-78010-4_16

Masking:	None (Open Label)
Primary Purpose:	Treatment
Official Title:	The Study of Human Epicardial Injection with Allogeneic Human-Induced Pluripotent Stem Cell-Derived Cardiomyocytes in Ischemic Heart Failure
Estimated Study Start Date:	May 1, 2019
Estimated Primary Completion Date:	December 1, 2020
Estimated Study Completion Date:	December 1, 2020

Arms and Interventions

Arm	Intervention/treatment
Experimental: hPSC-CM therapy Procedure: Injection of allogeneic human-induced pluripotent stem cell-derived cardiomyocytes (hiPSC-CMs) at the time of coronary artery bypass grafting surgery; 100 million hiPSC-CMs in 2.5–5-mL-medium suspension will be injected into the myocardium	Biological: hPSC-CM therapy Injection of allogeneic human-induced pluripotent stem cell-derived cardiomyocytes (hiPSC-CMs) at the time of coronary artery bypass grafting surgery; 100 million hiPSC-CMs in 2.5–5-mL medium suspension will be injected into the myocardium

Outcome Measures
Primary Outcome Measures:
1. The number of patients with serious adverse events associated with cell therapy after cardiomyocyte injection [time frame: 12 months]

   The primary safety endpoint of the study was the number of patients with serious adverse events associated with cell therapy after cardiomyocyte injection, primarily with persistent ventricular tachycardia, or accompanied by hemodynamic instability, or sudden cardiac death and examination of new tumors.

Secondary Outcome Measures:
1. Assessment of infarction size [time frame: at baseline, 3 months, 6 months, and 12 months]

   Measured using the nuclear magnetic delay enhancement technique

2. Change in local left ventricular function (cardiomyocyte injection area) [time frame: at baseline, 3 months, 6 months, and 12 months]

   Measured using the nuclear magnetic delay enhancement technique

3. Change in local left ventricular function (cardiomyocyte injection area) [time frame: at baseline, 3 months, 6 months, and 12 months]

   Measured via echocardiography

4. Change in local left ventricular thickness [time frame: at baseline, 3 months, 6 months, and 12 months]

   Measured using the nuclear magnetic delay enhancement technique

5. Change in local left ventricular thickness [time frame: at baseline, 3 months, 6 months, and 12 months]

   Measured via echocardiography

6. Change in local left ventricular thickness (at the end of diastole) [time frame: at baseline, 3 months, 6 months, and 12 months]

   Measured via echocardiography

7. Change in left ventricle ejection fraction (LVEF) [time frame: at baseline, 3 months, 6 months, and 12 months]

   Measured using the nuclear magnetic delay enhancement technique

8. Change in left ventricle ejection fraction (LVEF) [time frame: at baseline, 3 months, 6 months, and 12 months]

   Measured via echocardiography

9. Change in left ventricle end-diastolic volume (LVEDV) [time frame: at baseline, 3 months, 6 months, and 12 months]

   Measured using the nuclear magnetic delay enhancement technique

10. Change in left ventricle end-diastolic volume (LVEDV) [time frame: at baseline, 3 months, 6 months, and 12 months]

    Measured via echocardiography

11. Change in left ventricle end-systolic volume (LVESV) [time frame: at baseline, 3 months, 6 months, and 12 months]

    Measured using the nuclear magnetic delay enhancement technique

12. Change in left ventricle end-systolic volume (LVESV) [time frame: at baseline, 3 months, 6 months, and 12 months]

    Measured via echocardiography

13. Assessment of myocardial segmental perfusion [time frame: at baseline, 3 months, 6 months, and 12 months]

    Measured via cMRI

14. Cardiopulmonary function test: peak VO2 and PVO2 [time frame: at baseline, 3 months, 6 months]

Accurately measuring VO2 max involves a physical effort sufficient for the duration and intensity to fully tax the aerobic energy system. In general clinical and athletic testing, this usually involves a graded exercise test (either on a treadmill or on a cycle ergometer) in which the exercise intensity is progressively increased while measuring the ventilation and oxygen and carbon dioxide concentration of the inhaled and exhaled air. VO2 max is reached when oxygen consumption remains at a steady state despite an increase in workload.

15. Efficacy clinical function assessed by change in the 6-minute walking test [time frame: at baseline, 6 months]

A performance-based function that measures an individual's exercise capacity during a total of 6 minutes while walking on a hard, flat surface

16. New York Heart Association (NYHA) classification [time frame: at baseline and 3, 6, 9, and 12 months]

The cardiac function was categorized into four classes (I–IV).

17. Minnesota Living with Heart Failure Questionnaire (MHLFQ) score [time frame: at baseline and 3, 6, 9, and 12 months]

The minimum score is 0, and the maximum score is 105. And if the score change >= 10 scores improvement, change <=10 scores deteriorate, no meaningful change.

18. Incidence of the major adverse cardiac event (MACE) endpoint, defined as the composite incidence of (1) death, (2) hospitalization for heart failure, or (3) non-fatal recurrent myocardial infarction [time frame: 12 months]
19. Incidence of treatment-emergent adverse events [time frame: 12 months]
20. The incidence of arrhythmia [time frame: 48 hours after the procedure]

Holter monitoring for 48 hours will be used to monitor the heartbeat.

21. Concentration of hs-cTnT, cTnI, and CKMB [time frame: every 12 hours for the first 48 hours after CABG surgery]

These parameters are serial myocardial injury marker.

22. N-Terminal pro-brain natriuretic peptide (NT-pro-BNP) level [time frame: 12 months]

Eligibility Criteria

Ages Eligible for Study:	35 Years to 75 Years (Adult, Older Adult)
Sexes Eligible for Study:	All
Accepts Healthy Volunteers:	No

Inclusion Criteria:
1. Patients aged 35–75 years
2. Signed informed consent
3. Severe left ventricular systolic dysfunction with myocardial infarction history
4. New York Heart Association (NYHA) Class III or IV despite optimal standard of care
5. Indication for a conventional cardiac surgical procedure: coronary artery bypass grafting (CABG) (1) Class I (a). Left ventricular dysfunction with obvious left

main coronary artery lesion (level of evidence: B) (b). Left ventricular dysfunction with a similar left main coronary artery lesion such as more significant (>70% diameter) coronary artery stenoses proximal to the left anterior descending or left circumflex artery (level of evidence: B) (c). Left ventricular dysfunction with two or three lesions of the proximal left anterior descending (level of evidence: B) (2) Class II (a). Left ventricular dysfunction with a large number of surviving, non-contracting, or revascularizable cardiomyocytes, excluding the abovementioned coronary anatomy (level of evidence: B)

6. 30% < Ejection Fraction (LVEF) < 45% as assessed via echocardiography (measurement in the first 6 months of recruitment is included, excluding the measured values within 1 month of myocardial infarction)
7. Echocardiography indicates a segmental wall motion reduction or no motion

Exclusion Criteria:
1. Patients who have received treatments such as pacemakers or ICD or CRT device
2. Patients with severe valvular disease or presence of a mechanical valve replacement
3. Patients with persistent ventricular tachycardia or sudden cardiac death history
4. Baseline glomerular filtration rate <30 mL/min/1.73 m^2
5. Have a hematologic abnormality as evidenced by hematocrit <25%
6. Known, serious radiographic contrast allergy
7. Have a coagulopathy condition
8. Known allergies to penicillin or streptomycin
9. Contraindication to performance of a magnetic resonance imaging scan
10. Organ transplant recipient
11. History of malignant disease
12. Non-cardiac condition that limits lifespan to <1 year
13. On chronic therapy with immunosuppressant medication such as glucocorticoid or TNFα antagonist
14. Contraindication to take immunosuppressant medication
15. Serum-positive for HIV, hepatitis BsAg, or non-viremic hepatitis C
16. Participation in another intervention trial
17. Female patient who is pregnant or nursing

Part V
Diabetes Mellitus: Type 1 Diabetes (United States)

Chapter 17
Cellular Therapy for Type 1 Diabetes Using Mesenchymal Stem Cells

Brief Summary:

The goal of this study is to determine the safety and efficacy of fresh metabolically active allogeneic umbilical cord-derived mesenchymal stromal cells (UC-MSCs) for the treatment of new-onset type 1 diabetes (T1D) and to understand the mechanisms of protection. If proven effective, such a strategy can be used as a therapeutic option for T1D patients and potentially other autoimmune disorders.

Condition or disease	Intervention/treatment	Phase
Diabetes mellitus, type 1	Biological: Mesenchymal stem cells (MSCs) Other: Placebo infusion (Plasmalyte A with 0.5% human serum albumin)	Phase 1

Detailed Description:

This study seeks to find and enroll participants with new-onset type 1 diabetes (T1D) within 3 months of the first dose of insulin. T1D is an autoimmune disease in which T cells attack and destroy insulin-secreting pancreatic β cells, leading to insulin deficiency and hyperglycemia in patients. Life-long insulin therapy is the major treatment option. However, insulin therapy is not a cure, and thus a safer and more effective therapy is needed.

Mesenchymal stromal cells (MSCs) have emerged as a novel biopharmaceutical approach for many disorders. MSCs are a cellular product that can be derived from a patient's own body (autologous) or from a donor (allogeneic). This study will obtain MSCs from umbilical cords at the time of delivery from normal women who have been extensively screened for infectious diseases. These cells produced at the MUSC Center for Cellular therapy will be used within 3 passages after collection.

Medical University of South Carolina, Charleston, South Carolina, USA
ClinicalTrials.gov Identifier: NCT04061746

J. N. Weiss, *Stem Cell Surgery Trials in Heart Failure and Diabetes*,
https://doi.org/10.1007/978-3-030-78010-4_17

Evidence from animal models and clinical trials suggests that MSC infusion suppresses autoimmune and inflammatory diseases such as T1D. One clear message from these trials is that MSCs are effective at suppressing autoimmunity and seem generally safe. This study will measure the safety and efficacy of MSCs over the course of 1 year.

Study Type:	Interventional (Clinical Trial)
Estimated Enrollment:	50 Participants
Allocation:	Randomized

Actual Study Start Date:	February 13, 2020
Estimated Primary Completion Date:	July 1, 2023
Estimated Study Completion Date:	July 1, 2023

Arm	Intervention/treatment
Experimental: Group A Treatment 2.5×10^6 MSC per kg will be infused intravenously on Day 1	Biological: Mesenchymal stem cells (MSCs) Patients in Group A will receive a single MSC infusion
Placebo comparator: Group B Placebo Plasmalyte with 0.5% human serum albumin will be infused intravenously on Day 1	Other: Placebo Infusion (Plasmalyte A with 0.5% human serum albumin) Patients in Group B will receive a single infusion of placebo

Primary Outcome Measures:
1. 12-Month change in C-peptide area under the curve after a 2-hour MMTT [time frame: 1 year (plus or minus 30 days) after infusion]

   Change in beta cell function

Secondary Outcome Measures:
1. 6-Month change in C-peptide area under the curve after a 2-hour MMTT [time frame: 6 months (plus or minus 14 days) after infusion]

   Change in beta cell function

2. 6-Month peak C-peptide after a 2-hour MMTT [time frame: 6 months (plus or minus 14 days) after infusion]

   Change in beta cell function

3. 1-Year peak C-peptide after a 2-hour MMTT [time frame: 1 year (plus or minus 30 days) after infusion]

   Change in beta cell function

4. Change in 24-hour insulin dose per kilogram between baseline and 1 year measurements [time frame: 1 year (plus or minus 30 days) after infusion]

   Change in beta cell function

Other Outcome Measures:

1. Fasting and postprandial blood glucose levels after MSC infusion [time frame: 0–72 hours]

   Change in beta cell function

2. Changes in basal C-peptide and hemoglobin A1c [time frame: over the course of 1 year (0, 1, 3, 6, and 12 months)]

   Change in beta cell function

3. Change in serum glucagon levels [time frame: over the course of 1 year (0, 1, 3, 6, and 12 months)]

   Change in alpha cell function

4. Insulin secretion rate [time frame: over the course of 1 year (0, 1, 3, 6, and 12 months)]

   Change in beta cell function

5. Changes in islet autoantibodies [time frame: over the course of 1 year (0, 1, 3, 6, and 12 months)]

   Change in autoantibody presence or titer

6. Change in beta cell death measurements [time frame: over the course of 1 year (0, 1, 3, 6, and 12 months)]

   Determination of the mechanism of action

7. Change in blood T-reg number and function [time frame: over the course of 1 year (0, 1, 3, 6, and 12 months)]

   Determination of the mechanism of action

8. Change in autoantigen-specific T-cell response [time frame: over the course of 1 year (0, 1, 3, 6, and 12 months)]

   Determination of the mechanism of action

9. Change in blood autoreactive B cell number, B cell survival, and function [time frame: over the course of 1 year (0, 1, 3, 6, and 12 months)]

   Determination of the mechanism of action

10. Changes in mRNA expression in peripheral blood mononuclear cells after treatment [time frame: over the course of 1 year (0, 1, 3, 6, and 12 months)]

    Determination of the mechanism of action

11. Changes in serum cytokine levels after treatment [time frame: over the course of 1 year (0, 1, 3, 6, and 12 months)]

    Determination of the mechanism of action

Ages Eligible for Study:	12 Years to 30 Years (Child, Adult)
Sexes Eligible for Study:	All
Accepts Healthy Volunteers:	No

Criteria
Inclusion criteria:
- A new diagnosis of T1D based on the ADA criteria within 3 months of randomization
- Male and female between the ages of 12 and 30
- Mentally stable and able to comply with the procedures of the study protocol
- Positivity for at least one T1D-associated autoantibody, such as GAD, IA-2, or ZnT8 autoantibodies
- At screening, patients must have residual β-cell function with a stimulated peak C-peptide >0.2 nmol/L during a 2-hour MMTT
- Must be willing to comply with "intensive diabetes management" (*See diabetes management at MUSC below) as directed by the participant's clinician with the goal of maintaining blood glucose as close to normal as possible
- Subject must be willing to comply with the schedule of study visits and protocol requirements
- Subject with normal laboratory values as follows: white blood cell count, between 4500 and 11,000 per microliter; platelet counts, 140,000–450,000 platelets per microliter of blood; serum creatinine range, 0.6–1.3 mg/dL; hepatic function, ALT 5–55 units per liter (U/L) and AST 5–48 U/L

Exclusion criteria:
- Evidence of retinopathy at baseline based on ophthalmologic examination or medical record review
- Body mass index <14 or >35
- Presence of malignancy
- Subject has abnormally high lipid levels that exceed >3 times the upper limit of normal for LDL cholesterol or triglycerides
- Subject has blood pressure greater than 160 mmHg systolic or 100 mmHg diastolic at the time of consent
- Subject is being treated for severe active infection of any type
- A female subject who is breastfeeding, pregnant, or intends to become pregnant during the study
- Subject with clinically relevant uncontrolled medical condition not associated with diabetes (e.g., severe psychiatric, hematologic, renal, hepatic, neurologic, cardiac, or respiratory disorder)
- Subjects with HgbA1c >12% and/or fasting blood glucose >270 mg/dL and/or frequent episodes of hypoglycemia (>2 episodes per week of blood glucose levels <60 mg/dL)

Chapter 18
A Pilot Study of the Therapeutic Potential of Stem Cell Educator Therapy in Type 1 Diabetes

Brief Summary:
This is a prospective, single-arm, open-label, single-center pilot study to assess the safety, feasibility, and efficacy of stem cell educator therapy for the treatment of patients with Type 1 Diabetes.

Condition or disease	Intervention/treatment	Phase
Diabetes mellitus type 1	Genetic: Stem cell educator therapy	Phase 1

Detailed Description:
Our previous work demonstrated that human cord blood-derived multipotent stem cells (CB-SCs) are a unique type of stem cell identified from human cord blood, distinct from other stem cell types, including hematopoietic stem cells (HSCs) and mesenchymal stem cells (MSCs). The stem cells have harnessed some of their unique properties with stem cell educator therapy by using CB-SCs in a closed-loop system that circulates a patient's blood through a blood cell separator, briefly co-cultures the patient's lymphocytes with adherent CB-SCs in vitro, and returns the "educated" lymphocytes (but not the CB-SCs) to the patient's circulation. This treatment leads to global immune modulations and immune balance as demonstrated by clinical data and animal studies. The stem cell educator therapy may revolutionize the clinical treatment of diabetes and other immune-related diseases through CB-SCs' immune education and induction of immune balance without the safety and ethical concerns associated with conventional stem cell-based approaches

Hackensack University Medical Center – John Theurer Cancer Center, Hackensack, New Jersey, USA
ClinicalTrials.gov Identifier: NCT02624804

© The Author(s), under exclusive license to Springer Nature   89
Switzerland AG 2022
J. N. Weiss, *Stem Cell Surgery Trials in Heart Failure and Diabetes*,
https://doi.org/10.1007/978-3-030-78010-4_18

Study Type:	Interventional (Clinical Trial)
Estimated Enrollment:	10 Participants
Allocation:	N/A
Intervention Model:	Single-Group Assignment

Actual Study Start Date:	June 27, 2017
Estimated Primary Completion Date:	August 2020
Estimated Study Completion Date:	August 2022

Arm	Intervention/treatment
Experimental: Stem cell educator therapy Patients will have apheresis performed and then have their own blood returned to them with the "educated" lymphocytes	Genetic: Stem cell educator therapy patients will receive apheresis and then have their own blood returned to them with the "educated" lymphocytes Other name: stem cell education

Primary Outcome Measures:
1. Treatment-related adverse events [time frame: 12 months]

The primary study endpoint will be the occurrence of treatment-related adverse effects. Adverse events that occur during therapy (especially those that necessitate temporary or permanent discontinuation of therapy) and over the 12-month follow-up period will be assessed.

Secondary Outcome Measures:
1. Number of patients unable to complete therapy [time frame: 1 week]

Number of patients who were unable to complete the SCE therapy

Ages Eligible for Study:	18 Years and older (Adult, Older Adult)
Sexes Eligible for Study:	All
Accepts Healthy Volunteers:	No

Criteria
Inclusion Criteria:
- Adult patients >/=18 years
- Must have a diagnosis of type 1 diabetes mellitus based on the 2015 American Diabetes Association criteria for the Clarification and Diagnosis of diabetes
- Must have a blood test confirming the presence of at least one autoantibody to pancreatic islet cells (IAA, IA2, GAD 65, ZnT8)
- Fasting C-peptide level >0.3 ng/mL
- Adequate venous access for apheresis
- Ability to provide informed consent
- Must agree to comply with all study requirements and be willing to complete all study visits

Exclusion Criteria:
- AST or ALT 2> × upper limit of normal
- Creatinine >2.0 mg/dL
- Known coronary artery disease or EKG suggestive of coronary artery disease unless cardiac clearance for apheresis is obtained from a cardiologist
- Known active infection
- Pregnancy or breastfeeding mothers
- Use of immunosuppressive medication within 1 month of enrollment, including but not limited to prednisone, cyclosporine, tacrolimus, sirolimus, and chemotherapy
- Presence of any other autoimmune diseases (lupus, rheumatoid arthritis, scleroderma, etc.)
- Anticoagulation other than ASA
- Hemoglobin <10 g/dL or platelets <100 k/mL
- Is unable or unwilling to provide informed consent
- Presence of any other physical or psychological medical condition that, in the opinion of the investigator, would preclude participation

Chapter 19
Clinical Application of Stem Cell Educator Therapy in Type 1 Diabetes

Brief Summary:
Type 1 diabetes (T1D) is a T-cell-mediated autoimmune disease that causes a deficit in pancreatic islet beta cells. Millions of individuals worldwide have T1D, and its incidence increases annually. Several recent clinical trials point to the need for an approach that produces comprehensive immune modulation at both the local pancreatic and systemic levels. Stem cell educator (SCE) therapy offers comprehensive immune modulation at both the local and systemic levels in T1D by using a patient's own immune cells (including platelets) that are "educated" by cord blood stem cells. Tested clinically in more than 200 patients, SCE therapy has shown lasting reversal in autoimmunity in T1D patients, including improved C-peptide levels, reduced median glycated hemoglobin A1C (HbA1C) values, and decreased median daily usage of insulin. SCE therapy circulates a patient's blood through a blood cell separator, briefly co-cultures the patient's immune cells with adherent cord blood stem cells (CB-SCs) in vitro, and returns the "educated" autologous immune cells to the patient's circulation.

Condition or disease	Intervention/treatment	Phase
Type 1 diabetes	Combination product: Stem cell educator therapy	Phase 2 Phase 3

Detailed Description:
The SCE device is made of a hydrophobic material from FDA-approved (USP Class VI) dishes that tightly binds CB-SCs without interfering with their immune-modulating capability. We originally designed a chamber for co-culture of lymphocytes and CB-SCs that included nine discs of the material with a flow pathway and

Hackensack Meridian Health, Hackensack, New Jersey, USA
ClinicalTrials.gov Identifier: NCT04011020

adherent CB-SCs sandwiched between a top cover plate and a bottom collecting plate. In this trial, we are going to use the 12-layer SCE device.

The SCE therapy carried a lower risk of infection than a typical blood transfusion and did not introduce stem cells or reagents into the patients. In addition, CB-SCs have very low immunogenicity, and the CB-SCs cultured in the device are a highly restricted population and contain no CD3+ T cells or other lymphocyte subsets, eliminating the need for human leukocyte antigen (HLA) matching prior to treatment. This innovative approach has the potential to provide CB-SC-mediated immune modulation therapy for multiple autoimmune diseases while mitigating the safety and ethical concerns associated with other approaches, such as T1D, type 2 diabetes (T2D), and alopecia areata (AA), in clinics. The relative simplicity of the approach may also provide cost and time savings relative to other approaches.

Study Type:	Interventional (Clinical Trial)
Estimated Enrollment:	50 Participants
Allocation:	N/A
Intervention Model:	Single-Group Assignment

Arm	Intervention/treatment
Experimental: Treatment of T1D with stem cell educator therapy Recruited T1D subjects will receive one treatment with SCE therapy	Combination product: Stem cell educator therapy Patients with T1D will be evaluated by the study principal investigator or co-investigators. Informed consent will be obtained at the initial screening visit. The initial screening visit will occur within 30 days of initiation of the SCE therapy. The second screening visit will occur within 7 days of therapy. Subjects who meet all criteria will be scheduled for treatment. All enrolled subjects will receive treatment with the SCE system consisting of a single session of mononuclear cell (MNC) collection by apheresis where 10 L of blood will be processed on day −1. The MNC product will then be exposed to the SCE, and on day 0, the product will be infused intravenously back to the patient.

Primary Outcome Measures:
1. Incidence of treatment adverse events in T1D subjects [time frame: 6 months]

The occurrence of treatment-related adverse events will be evaluated post the treatment with the SCE therapy

Secondary Outcome Measures:
1. Preliminary efficacy of SCE therapy to improve beta cell function [time frame: 12 months]

Preliminary efficacy as measured by area under the C-peptide curve (AUC) over the first 2 hours of a 3-hour mixed meal tolerance test (MMTT)

2. Preliminary efficacy of SCE therapy to improve glucose control [time frame: 12 months]

Change in HbA1C levels over time

3. Preliminary efficacy of SCE therapy to reduce insulin dose [time frame: 12 months]

Change in daily insulin requirements

4. Efficacy of SCE therapy in immune modulation [time frame: 12 months]

Measurements of immune markers at baseline, 1, 3, 6, 9, and 12 months. Peripheral blood mononuclear cells (PBMC) will be collected and tested by flow cytometry.

Ages Eligible for Study:	12 Years and older (Child, Adult, Older Adult)
Sexes Eligible for Study:	All
Accepts Healthy Volunteers:	No

Criteria
Inclusion Criteria:
1. Adult patients (12 years)
2. Must have a diagnosis of type 1 diabetes mellitus based on the 2015 American Diabetes Association criteria for the Clarification and Diagnosis of diabetes
3. Must have a blood test confirming the presence of at least one autoantibody to pancreatic islet cells (IAA, IA2, GAD 65, ZnT8)
4. Fasting C-peptide level >0.3 ng/mL
5. HbA1C <10% at enrollment
6. Recent diagnosis (within 2 years of enrollment)
7. Adequate venous access for apheresis
8. Must be equipped with a continuous glucose monitoring system (CGMS)
9. Ability to provide informed consent
10. For female patients only, willingness to use FDA-recommended birth control (http://www.fda.gov/downloads/ForConsumers/ByAudience/ForWomen/FreePublications/UCM356451.pdf) until 6 months post treatment
11. Must agree to comply with all study requirements and be willing to complete all study visits

Exclusion Criteria:
1. AST or ALT 2> × upper limit of normal
2. Abnormal bilirubin (total bilirubin >1.2 mg/dL, direct bilirubin >0.4 mg/dL)
3. Creatinine > 2.0 mg/dL
4. Known coronary artery disease or EKG suggestive of coronary artery disease unless cardiac clearance for apheresis is obtained from a cardiologist
5. Known active infection such as hepatitis B, hepatitis C, or human immunodeficiency virus (HIV)
6. Pregnancy assessed by a positive serum pregnancy test or breastfeeding mothers

7. Use of immunosuppressive medication within 1 month of enrollment including but not limited to prednisone, cyclosporine, tacrolimus, sirolimus, and chemotherapy
8. Presence of any other autoimmune diseases (lupus, rheumatoid arthritis, scleroderma, etc.)
9. Anticoagulation other than ASA
10. Hemoglobin <10 g/dL or platelets <100 k/mL
11. Is unable or unwilling to provide informed consent
12. Presence of any other physical or psychological medical condition that, in the opinion of the investigator, would preclude participation

Publications of Results:
Delgado E, Perez-Basterrechea M, Suarez-Alvarez B, Zhou H, Revuelta EM, Garcia-Gala JM, Perez S, Alvarez-Viejo M, Menendez E, Lopez-Larrea C, Tang R, Zhu Z, Hu W, Moss T, Guindi E, Otero J, Zhao Y. Modulation of Autoimmune T-Cell Memory by Stem Cell Educator Therapy: Phase 1/2 Clinical Trial. EBioMedicine. 2015;2(12):2024-36. https://doi.org/10.1016/j.ebiom.2015.11.003. eCollection 2015 Dec.

Zhao Y, Jiang Z, Zhao T, Ye M, Hu C, Yin Z, Li H, Zhang Y, Diao Y, Li Y, Chen Y, Sun X, Fisk MB, Skidgel R, Holterman M, Prabhakar B, Mazzone T. Reversal of type 1 diabetes via islet β cell regeneration following immune modulation by cord blood-derived multipotent stem cells. BMC Med. 2012;10:3. https://doi.org/10.1186/1741-7015-10-3.

Zhao Y. Stem cell educator therapy and induction of immune balance. Curr Diab Rep. 2012;12(5):517–23. https://doi.org/10.1007/s11892-012-0308-1. Review.

Zhao Y, Jiang Z, Delgado E, Li H, Zhou H, Hu W, Perez-Basterrechea M, Janostakova A, Tan Q, Wang J, Mao M, Yin Z, Zhang Y, Li Y, Li Q, Zhou J, Li Y, Martinez Revuelta E, Maria García-Gala J, Wang H, Perez-Lopez S, Alvarez-Viejo M, Menendez E, Moss T, Guindi E, Otero J. Platelet-Derived Mitochondria Display Embryonic Stem Cell Markers and Improve Pancreatic Islet β-cell Function in Humans. Stem Cells Transl Med. 2017;6(8):1684–97. https://doi.org/10.1002/sctm.17-0078. Epub 2017 Jul 7.

Chapter 20
Patient-Derived Stem Cell Therapy for Diabetic Kidney Disease

Brief Summary:
The researchers will assess the safety, tolerability, dosing effect, and early signals of efficacy of intra-arterially delivered autologous (from self) adipose (fat) tissue-derived mesenchymal stem/stromal cells (MSC) in patients with progressive diabetic kidney disease (DKD).

Condition or disease	Intervention/treatment	Phase
Diabetic kidney disease Diabetic nephropathies Diabetes mellitus, type 2 Diabetes mellitus, type 1 Chronic kidney disease Diabetic nephropathy type 2 Kidney failure Kidney insufficiency	Biological: Autologous adipose-derived mesenchymal stem/stromal cells (MSC) lower dose Biological: Autologous adipose-derived mesenchymal stem/stromal cells (MSC) higher dose	Phase 1

Detailed Description:
This is a single-center, open-label, dose-escalating study assessing safety, tolerability, dosing effect, and early signals of efficacy of intra-arterially delivered autologous (from self) adipose tissue-derived mesenchymal stem/stromal cells (MSCs) in 30 patients with progressive diabetic kidney disease (DKD). DKD will be defined as chronic kidney disease (CKD; estimated glomerular filtration rate (eGFR) <60 mL/min/1.73 m^2) in the setting of diabetes mellitus (type 2; on anti-diabetes therapy) without overt etiologies of CKD beyond concomitant hypertension.

Progressive DKD will be considered as eGFR 25–55 mL/min/1.73 m^2 with (a) eGFR decline of 5 mL/min over 18 months or 10 mL/min over 3 years or (b) an

Mayo Clinic in Rochester, Rochester, Minnesota, USA
ClinicalTrials.gov Identifier: NCT03840343

© The Author(s), under exclusive license to Springer Nature Switzerland AG 2022
J. N. Weiss, *Stem Cell Surgery Trials in Heart Failure and Diabetes*, https://doi.org/10.1007/978-3-030-78010-4_20

intermediate or high 5-year risk of progression to end-stage kidney failure (dialysis or transplant) based on the validated Tangri 4-variable (age, sex, eGFR, urinary albumin-creatinine ratio) kidney failure risk equation. Fifteen subjects will be placed in one of two cell dosage arms in a parallel design with single-kidney MSC administration at Day 0 and Month 3. Subjects will be followed up for a total of 15 months from the time of initial cell administration.

Study Type:	Interventional (Clinical Trial)
Estimated Enrollment:	30 Participants
Allocation:	Non-randomized
Intervention Model:	Parallel Assignment

Actual Study Start Date:	October 23, 2019
Estimated Primary Completion Date:	December 2024
Estimated Study Completion Date:	December 2025

Arm	Intervention/treatment
Experimental: Lower-dose MSC This arm will receive autologous adipose-derived mesenchymal stem/stromal cell (MSC) lower dose.	Biological: Autologous adipose-derived mesenchymal stem/stromal cell (MSC) lower dose Two MSC infusions of 2.5×10^5 cells/kg at time zero and 3 months; single kidney, intra-arterial delivery
Experimental: Higher-dose MSC This arm will receive autologous adipose-derived mesenchymal stem/stromal cell (MSC) higher dose	Biological: Autologous adipose-derived mesenchymal stem/stromal cell (MSC) higher dose Two MSC infusions of 5.0×10^5 cells/kg at time zero and 3 months; single kidney, intra-arterial delivery

Primary Outcome Measures:
1. Adverse Events [time frame: baseline through Month 15]

   The number of adverse events associated with MSC intervention per treatment arm

2. Adverse events [time frame: baseline through Month 15]

   The percentage of adverse events associated with MSC intervention per treatment arm

Secondary Outcome Measures:
1. Kidney function [time frame: baseline, Month 6]

   Change in measured glomerular filtration rate (mGFR). Measured as mL/min/BSA

2. Kidney function [time frame: pretreatment, Month 12]

   Change in estimated glomerular filtration rate (eGFR) slope. Measured as mL/min/1.73 m²/month

Ages Eligible for Study:	45 Years to 75 Years (Adult, Older Adult)
Sexes Eligible for Study:	All
Accepts Healthy Volunteers:	No

Criteria
Inclusion Criteria:
- Diabetes mellitus (on anti-diabetes drug therapy)
- Age 45–75 years
- eGFR 25–55 mL/min/1.73 m² at the time of consent with (a) eGFR decline of 5 mL/min over 18 months or 10 mL/min over 3 years or (b) an intermediate or high 5-year risk of progression to end-stage kidney failure (dialysis or transplant) based on the validated Tangri 4-variable (age, sex, eGFR, urinary albumin–creatinine ratio) kidney failure risk equation (https://kidneyfailurerisk.com/)
- The primary cause of kidney disease is diabetes without suspicion of concomitant kidney disease beyond hypertension
- Spot urine albumin–creatinine ≥30 mg/g unless on RAAS inhibition
- Ability to give informed consent

Exclusion Criteria:
- Hemoglobin A1c ≥11%
- Pregnancy
- Active malignancy
- Active immunosuppression therapy
- Kidney transplantation history
- Concomitant glomerulonephritis
- Nephrotic syndrome
- Solid organ transplantation history
- Autosomal dominant or recessive polycystic kidney disease
- Known renovascular disease
- Kidney failure (hemodialysis, peritoneal dialysis, or kidney transplantation)
- Active tobacco use
- Body weight >150 kg or BMI >50
- Uncontrolled hypertension: systolic blood pressure (SBP) >180 mmHg despite antihypertensive therapy
- Recent cardiovascular event (myocardial infarction, stroke, congestive heart failure within 6 months)
- Evidence of hepatitis B or C or HIV infection, chronic
- Anticoagulation therapy requiring heparin bridging for procedures
- History of methicillin-resistant *Staphylococcus aureus* colonization
- Recent plastic, chemical, or surgical manipulation of adipose tissue for cosmetic purposes within 6 months
- Inability to give informed consent
- Potentially unreliable subjects and those judged by the investigator to be unsuitable for the study

Part VI
Diabetes Mellitus: Type 1 Diabetes (China)

Chapter 21
Stem Cell Educator Therapy in Diabetes

Brief Summary:
Stem cell educator (SCE) therapy circulates a patient's blood through a blood cell separator, briefly co-cultures the patient's immune cells with adherent cord blood stem cells (CB-SCs) in vitro, and returns only the "educated" autologous immune cells to the patient's circulation. Several mechanistic studies with clinical samples and animal models have demonstrated the proof of concept and clinical safety of SCE therapy. They suggest that SCE therapy may function via CB-SC induction of immune tolerance in the autoimmune T cells and pathogenic monocytes/macrophages when these are exposed to the autoimmune regulator protein (AIRE) in the CB-SCs. In this project, the optimized SCE therapy for type 1 diabetes (T1D) and T2D will be tested in a prospective, single-arm, open-label, single-center study to assess its clinical efficacy and the related molecular mechanisms in patients with diabetes.

Condition or disease	Intervention/treatment	Phase
Diabetes mellitus, type 1 Diabetes mellitus, type 2	Combination product: Stem cell educator therapy	Phase 2

Study Type:	Interventional (Clinical Trial)
Estimated Enrollment:	100 Participants
Allocation:	N/A
Intervention Model:	Single-Group Assignment
Intervention Model Description:	In this project, the optimized SCE therapy for diabetes will be tested in a prospective, single-arm, open-label, single-center study

Department of Endocrinology, Chinese PLA General Hospital, Beijing, China
ClinicalTrials.gov Identifier: NCT03390231

Actual Study Start Date:	November 27, 2017
Estimated Primary Completion Date:	December 30, 2020
Estimated Study Completion Date:	December 31, 2020

Arm	Intervention/treatment
Experimental: Stem cell educator	Combination product: Stem cell
The stem cell educator (SCE) technology involves a closed-loop system that circulates a patient's blood through a blood cell separator, briefly co-cultures the patient's immune cells with adherent CB-SCs in vitro, and returns only the "educated" immune cells to the patient's circulation. Several mechanistic studies with clinical samples and animal models have been conducted to demonstrate the proof of concept and clinical safety of SCE therapy. They suggest that SCE therapy may function via CB-SC induction of immune tolerance in the autoimmune T cells and pathogenic monocytes/macrophages that are encountered through the action of the autoimmune regulator (AIRE) and other molecular mechanisms. After the induction of immune tolerance in the immune cells, the immune balance and homeostasis may be restored when treated cells are returned in vivo.	educator therapy It briefly co-cultures the patient's lymphocytes with CB-SCs in vitro, induces immune tolerance through the action of autoimmune regulator (AIRE, expressed by CB-SCs), returns the educated autologous lymphocytes to the patient's circulation, and restores immune balance and homeostasis

Primary Outcome Measures:
1. Changes in inflammation-related markers in diabetic patients after stem cell educator therapy [time frame: 30 days]

After treatment for 30 days, diabetic patients will be tested for inflammation-related markers (e.g., Th1/Th2 cytokines) by flow cytometry, which will be compared with the baseline levels.

Secondary Outcome Measures:
1. Change in insulin resistance [time frame: 30 days]

Before treatment, test for insulin sensitivity by chip analysis as baseline; after treatment, test sensitivity levels on the 1st month.

2. Metabolic control in HbA1C levels [time frame: 3 months]

Before treatment, test for HbA1C levels as baseline; after treatment, repeat these testing on the 3rd month. Hemoglobin A1c (HbA1c) will be reported as the changes in percentage.

3. Metabolic control in blood glucose levels [time frame: 3 months]

Before treatment, test for fasting blood glucose as baseline; after treatment, repeat this testing on the 3rd month. The blood glucose level will be measured as mmol/L.

4. Metabolic control in fasting C-peptide levels [time frame: 3 months]

Before treatment, test for C-peptide levels as baseline; after treatment, repeat this testing on the 3rd month. C-peptide will be measured as ng/mL.

Ages Eligible for Study:	20 Years to 60 Years (Adult)
Sexes Eligible for Study:	All
Accepts Healthy Volunteers:	No

Criteria
Inclusion Criteria:
- Patients are screened for enrollment in the study if both clinical signs and laboratory tests meet the diagnosis standards of the American Diabetes Association.

Exclusion Criteria:
- Exclusion criteria are any clinically significant diseases in the liver, kidney, and heart. Additional exclusion criteria are pregnancy, immunosuppressive medication, viral diseases, or diseases associated with immunodeficiency.

Publications of Results:
Delgado E, Perez-Basterrechea M, Suarez-Alvarez B, Zhou H, Revuelta EM, Garcia-Gala JM, Perez S, Alvarez-Viejo M, Menendez E, Lopez-Larrea C, Tang R, Zhu Z, Hu W, Moss T, Guindi E, Otero J, Zhao Y. Modulation of autoimmune T-cell memory by stem cell educator therapy: phase 1/2 clinical trial. EBioMedicine. 2015;2(12):2024–36.   https://doi.org/10.1016/j.ebiom.2015.11.003.   eCollection 2015.

Zhao Y, Jiang Z, Delgado E, Li H, Zhou H, Hu W, Perez-Basterrechea M, Janostakova A, Tan Q, Wang J, Mao M, Yin Z, Zhang Y, Li Y, Li Q, Zhou J, Li Y, Martinez Revuelta E, Maria García-Gala J, Wang H, Perez-Lopez S, Alvarez-Viejo M, Menendez E, Moss T, Guindi E, Otero J. Platelet-derived mitochondria display embryonic stem cell markers and improve pancreatic islet β-cell function in humans. Stem Cells Transl Med. 2017;6(8):1684–1697. https://doi.org/10.1002/sctm.17-0078.

Zhao Y, Jiang Z, Zhao T, Ye M, Hu C, Yin Z, Li H, Zhang Y, Diao Y, Li Y, Chen Y, Sun X, Fisk MB, Skidgel R, Holterman M, Prabhakar B, Mazzone T. Reversal of type 1 diabetes via islet β cell regeneration following immune modulation by cord blood-derived multipotent stem cells. BMC Med. 2012;10:3. https://doi.org/10.1186/1741-7015-10-3

Zhao Y, Jiang Z, Zhao T, Ye M, Hu C, Zhou H, Yin Z, Chen Y, Zhang Y, Wang S, Shen J, Thaker H, Jain S, Li Y, Diao Y, Chen Y, Sun X, Fisk MB, Li H. Targeting insulin resistance in type 2 diabetes via immune modulation of cord blood-derived multipotent stem cells (CB-SCs) in stem cell educator therapy: phase I/II clinical trial. BMC Med. 2013;11:160. https://doi.org/10.1186/1741-7015-11-160

Chapter 22
Reversal of Type 1 Diabetes in Children by Stem Cell Educator Therapy

Brief Summary:
Type 1 diabetes (T1D) is an autoimmune disease that usually occurs in children and reduces their pancreatic islet beta cells and thereby limits insulin production. Millions of individuals worldwide have T1D, and the number of children with diagnosed or undiagnosed T1D is increasing annually. Insulin supplementation is not a cure. It does not halt the persistent autoimmune response nor can it reliably prevent devastating complications such as neuronal and cardiovascular diseases, blindness, and kidney failure. A true cure has proven elusive despite intensive research pressure over the past 25 years. Notably, Dr. Zhao and his team have successfully developed a groundbreaking technology, namely, stem cell educator therapy (Zhao Y, et al. BMC Medicine 2011, 2012). To date, clinical trials in adult patients have demonstrated the safety and efficacy of stem cell educator therapy for the treatment of T1D and other autoimmune-associated diseases. Here, the investigators will evaluate the safety and efficacy of stem cell educator therapy in children with type 1 diabetes.

Condition or disease	Intervention/treatment	Phase
Type 1 diabetes	Device: Stem cell educator	Phase 1 Phase 2

Study Type:	Interventional (Clinical Trial)
Estimated Enrollment:	20 Participants
Allocation:	Randomized
Intervention Model:	Single-Group Assignment

The Second Xiangya Hospital, Changsha, Hunan, China
ClinicalTrials.gov Identifier: NCT01996228

© The Author(s), under exclusive license to Springer Nature
Switzerland AG 2022
J. N. Weiss, *Stem Cell Surgery Trials in Heart Failure and Diabetes*,
https://doi.org/10.1007/978-3-030-78010-4_22

Study Start Date:	November 2013
Estimated Primary Completion Date:	October 2019
Estimated Study Completion Date:	October 2019

Arm	Intervention/treatment

Experimental: Cord blood-derived multipotent stem cells
Human cord blood-derived multipotent stem cells (CB-SC) display unique phenotypes, such as the expression of embryonic stem (ES) cell markers, multipotential differentiations, very low immunogenicity, and immune modulations in patients
	Device: Stem cell educator
	Other names:
	  Procedure: Apheresis and stem cell educator therapy
	  Biological: Cord blood

Primary Outcome Measures:
1. Autoimmune control [time frame: 90 days post treatment]

Before treatment, test autoimmune-related markers as baseline; after treatment for 90 days, repeat testing of autoimmune-related markers.

Secondary Outcome Measures:
1. Metabolic control [time frame: 3–24 months post treatment]

Before treatment, test for C-peptide levels and HbA1C as baseline; after treatment, test C-peptide levels and HbA1C on the 3rd month:

(a) Analysis of islet beta cell function
(b) Test for C-peptide levels every 6 months
(c) Full evaluation of islet beta cell function after 2 years

Ages Eligible for Study:	6 Years to 14 Years (Child)
Sexes Eligible for Study:	All
Accepts Healthy Volunteers:	No

Criteria
Inclusion Criteria:
1. T1D patients are screened for enrollment in the study if both clinical signs and laboratory tests meet the diagnosis standards of the American Diabetes Association.
2. Children from 3 through 18 years old and body weight >15 kg.
3. Presence of at least one autoantibody to the pancreatic islet β cells (IA-2, GAD, ICA, ZnT8, or IAA).
4. Written informed consent from the child and child's parents or legal representative.

Exclusion Criteria:
1. Any clinically significant diseases in the liver, kidney, and heart
2. Additional exclusion criteria include immunosuppressive medication, viral diseases, or diseases associated with immunodeficiency
3. Significantly abnormal hematology results at screening
4. Presence of any infectious diseases or inflammation conditions, including active skin infections, flu, fever, upper or lower respiratory tract infections.

Publications:
Zhao Y, Jiang Z, Zhao T, Ye M, Hu C, Yin Z, Li H, Zhang Y, Diao Y, Li Y, Chen Y, Sun X, Fisk MB, Skidgel R, Holterman M, Prabhakar B, Mazzone T. Reversal of type 1 diabetes via islet β cell regeneration following immune modulation by cord blood-derived multipotent stem cells. BMC Med. 2012;10:3. https://doi.org/10.1186/1741-7015-10-3

Zhao Y, Jiang Z, Zhao T, Ye M, Hu C, Zhou H, Yin Z, Chen Y, Zhang Y, Wang S, Shen J, Thaker H, Jain S, Li Y, Diao Y, Chen Y, Sun X, Fisk MB, Li H. Targeting insulin resistance in type 2 diabetes via immune modulation of cord blood-derived multipotent stem cells (CB-SCs) in stem cell educator therapy: phase I/II clinical trial. BMC Med. 2013;11:160. https://doi.org/10.1186/1741-7015-11-160

Zhao Y, Mazzone T. Human cord blood stem cells and the journey to a cure for type 1 diabetes. Autoimmun Rev. 2010;10(2):103–7. https://doi.org/10.1016/j.autrev.2010.08.011. Epub 2010 Aug 20. Review.

Zhao Y. Stem cell educator therapy and induction of immune balance. Curr Diab Rep. 2012;12(5):517–23. https://doi.org/10.1007/s11892-012-0308-1. Review.

Part VII
Diabetes Mellitus: Type 1 Diabetes (China/Spain)

Chapter 23
Stem Cell Educator Therapy in Type 1 Diabetes

Brief Summary:
The translational potential to the clinical applications of cord blood stem cells has increased enormously in recent years, mainly because of its unique advantages, including no risk to the donor, no ethical issues, low risk of graft-versus-host disease (GVHD), rapid availability, and large resource worldwide. Human cord blood contains several types of stem cells such as the umbilical cord blood-derived multipotent stem cells (CB-SC). CB-SC possess multiple biological properties, including the expression of embryonic stem (ES) cell characteristics, giving rise to different types of cells and immune modulation. Specifically, CB-SC can function as an immune modulator that can lead to the control of the immune responses, which could in turn be used as a new approach to overcome the autoimmunity of type 1 diabetes (T1D) in patients and non-obese diabetic (NOD) mice. Here, the investigators develop a novel stem cell educator therapy by using CB-SC and explore the therapeutic effectiveness of educator therapy in T1D patients.

Condition or disease	Intervention/treatment	Phase
Type 1 diabetes	Device: Stem cell educator	Phase 2

Study Type:	Interventional (Clinical Trial)
Estimated Enrollment:	100 Participants
Allocation:	N/A
Intervention Model:	Single-Group Assignment

The First Hospital of Hebei Medical University, Shijiazhuang, Hebei, China
The Second Xiangya Hospital of Central South University, Changsha, Hunan, China
General Hospital of Jinan Military Command, Jinan, Shandong, China
Hospital Universitario Central de Asturias, Oviedo, Asturias, Spain
ClinicalTrials.gov Identifier: NCT01350219

© The Author(s), under exclusive license to Springer Nature Switzerland AG 2022
J. N. Weiss, *Stem Cell Surgery Trials in Heart Failure and Diabetes*,
https://doi.org/10.1007/978-3-030-78010-4_23

Study Start Date:	September 2010
Estimated Primary Completion Date:	September 2019
Estimated Study Completion Date:	September 2019

Arm	Intervention/treatment
Experimental: Cord blood stem cell Human cord blood-derived multipotent stem cells (CB-SC) display unique phenotypes, such as the expression of embryonic stem (ES) cell markers, multipotential differentiations, very low immunogenicity, and immune modulations	Device: Stem cell educator For the treatment, commonly the left (or right) median cubital vein, a patient's blood is passed through a blood cell separator that isolates the lymphocytes from the blood according to the recommended protocol by the manufacturer; consequently, the collected lymphocytes are transferred to the stem cell educator and treated by CB-SC; after that, the educated cells return the blood back to the patient via the dorsal vein of the hand. During the MCS+ collection, the whole blood flow rate is maintained at 35 mL/min. The whole procedure is scheduled for 8–9 h.

Primary Outcome Measures:
1. Autoimmune control [time frame: 30 days post treatment]

Before treatment, test autoimmune-related markers as baseline; after treatment for 30 days, repeat testing of autoimmune-related markers.

Secondary Outcome Measures:
1. Metabolic control [time Frame: 3 months]

Before treatment, test for C-peptide levels as baseline; after treatment, test C-peptide levels on the 3rd month.

2. Analysis of islet beta cell function [time frame: 6 months]

    (a) Test for C-peptide levels on the 6th month
    (b) Full evaluation of islet beta cell function after 1 year

Ages Eligible for Study:	14 Years to 60 Years (Child, Adult)
Sexes Eligible for Study:	All
Accepts Healthy Volunteers:	Yes

Criteria
Inclusion Criteria:
- Patients were screened for enrollment in the study if both clinical signs and laboratory tests meet the diagnosis standards of the American Diabetes Association 2010. Other key inclusion criteria were presence of at least one autoantibody to the pancreatic islet β cells.

Exclusion Criteria:

*   Exclusion criteria were any clinically significant diseases in the liver, kidney, and heart. Additional exclusion criteria were pregnancy, immunosuppressive medication, viral diseases, or diseases associated with immunodeficiency.

Publications:

Delgado E, Perez-Basterrechea M, Suarez-Alvarez B, Zhou H, Revuelta EM, Garcia-Gala JM, Perez S, Alvarez-Viejo M, Menendez E, Lopez-Larrea C, Tang R, Zhu Z, Hu W, Moss T, Guindi E, Otero J, Zhao Y. Modulation of autoimmune T-cell memory by stem cell educator therapy: phase 1/2 clinical trial. EBioMedicine. 2015;2(12):2024–36. https://doi.org/10.1016/j.ebiom.2015.11.003. eCollection 2015.

Zhao Y, Jiang Z, Zhao T, Ye M, Hu C, Yin Z, Li H, Zhang Y, Diao Y, Li Y, Chen Y, Sun X, Fisk MB, Skidgel R, Holterman M, Prabhakar B, Mazzone T. Reversal of type 1 diabetes via islet β cell regeneration following immune modulation by cord blood-derived multipotent stem cells. BMC Med. 2012;10:3. https://doi.org/10.1186/1741-7015-10-3

Zhao Y, Lin B, Darflinger R, Zhang Y, Holterman MJ, Skidgel RA. Human cord blood stem cell-modulated regulatory T lymphocytes reverse the autoimmune-caused type 1 diabetes in nonobese diabetic (NOD) mice. PLoS One. 2009;4(1):e4226. https://doi.org/10.1371/journal.pone.0004226. Epub 2009 Jan 19.

Zhao Y, Mazzone T. Human cord blood stem cells and the journey to a cure for type 1 diabetes. Autoimmun Rev. 2010;10(2):103–7. https://doi.org/10.1016/j.autrev.2010.08.011. Epub 2010 Aug 20. Review.

Zhao Y. Stem cell educator therapy and induction of immune balance. Curr Diab Rep. 2012;12(5):517–23. https://doi.org/10.1007/s11892-012-0308-1. Review. Publications automatically indexed to this study by ClinicalTrials.gov Identifier (NCT Number).

Part VIII
Diabetes Mellitus: Type 1 Diabetes
(South America)

Chapter 24
Mesenchymal Stem Cells in Patients with Type 1 Diabetes Mellitus

Brief Summary:
In this prospective, dual-center, open-label trial, patients with recent-onset type 1 diabetes will receive one dose of allogeneic adipose tissue-derived stromal/stem cells (1 × 106 cells/kg) and oral cholecalciferol 2000 UI/day for 24 months (group 1). They will be compared with patients that will receive just oral cholecalciferol 2000 UI/day (group 2) and standard treatment (group 3: no treatment). Adverse events will be recorded. In addition, glycated hemoglobin, insulin dose, frequency of hypoglycemia, glycemic variability, % of time in hyper and hypoglycemia, and peak response of the C-peptide after the mixed meal test will be measured at baseline (T0) and after 3 (T3), 6 (T6), 12 (T12), 18 (T18), and 24 (T24) months.

Condition or disease	Intervention/treatment	Phase
Type 1 diabetes mellitus	Biological: Infusion of adipose tissue-derived stem/stromal cells and oral cholecalciferol supplementation	Not applicable

Study Type:	Interventional (Clinical Trial)
Estimated Enrollment:	30 Participants
Allocation:	Randomized
Intervention Model:	Parallel Assignment

Actual Study Start Date:	March 1, 2015
Actual Primary Completion Date:	March 1, 2020
Estimated Study Completion Date:	December 1, 2025

Clementino Fraga Filho University Hospital of Rio de Janeiro, Rio de Janeiro, Brazil
ClinicalTrials.gov Identifier: NCT03920397

Arm	Intervention/treatment
Experimental: Adipose tissue-derived stem/stromal cells Safety of adipose tissue-derived stem/stromal cells (ASCs) for 24 months in patients with recent-onset type 1 diabetes	Biological: Infusion of adipose tissue-derived stem/stromal cells and oral cholecalciferol supplementation. The investigators will assess the area under the curve of C-peptide after a liquid mixed meal (Glucerna®) considering the time 0 (baseline), 30, 60, 90, and 120 min in each follow-up outpatient visit (3 [T3], 6 [T6], 12 [T12], 18 [T18], and 24 [T24] months after the adipose tissue-derived stem/stromal cells infusion). Other clinical and pancreatic function evaluation that will be assessed are weight, height, body mass index (BMI), blood pressure, heart frequency, frequency of hypoglycemia and insulin dose/kg of body weight, blood count, lipids, renal and hepatic function, thyroid-stimulating hormone, free thyroxine, antithyroglobulin antibody, calcium, phosphorus, and 25-hydroxy vitamin D
Experimental: Daily 2000 UI of oral cholecalciferol To investigate the efficacy of daily 2000 UI cholecalciferol for 24 months in patients with recent-onset type 1 diabetes	Biological: Infusion of adipose tissue-derived stem/stromal cells and oral cholecalciferol supplementation. The investigators will assess the area under the curve of C-peptide after a liquid mixed meal (Glucerna®), considering the time 0 (basal), 30, 60, 90, and 120 min in each follow-up outpatient visit (3 [T3], 6 [T6], 12 [T12], 18 [T18], and 24 [T24] months after the adipose tissue-derived stem/stromal cell infusion). Other clinical and pancreatic function evaluation that will be assessed are weight, height, body mass index (BMI), blood pressure, heart frequency, frequency of hypoglycemia and insulin dose/kg of body weight, blood count, lipids, renal and hepatic function, thyroid-stimulating hormone, free tyroxine, antithyroglobulin antibody, calcium, phosphorus and 25-hydroxy vitamin D

Primary Outcome Measures:

1. Pancreatic β-cell function after an adipose tissue-derived stem/stromal cell infusion [time frame: 24 months]

Thirty patients with recent-onset type 1 diabetes will receive adipose tissue-derived stem/stromal cells (ASCs). The patients will be admitted on the day of the infusion and will be discharged 24 h after infusion. A single dose of ASCs will be infused through the peripheral upper arm vein in 15–20 min.

Pancreatic β-cell function will be assessed at each follow-up outpatient visit (1, 3, 6, 12, 18, and 24 months after the ASC infusion). In each visit, C-peptide will be analyzed by immunofluorometric assay, considering the time 0 (basal), and peak stimulated C-peptide (30, 60, 90, and 120 min) after the mixed meal test (Glucerna)

2. Glycemic control after an adipose tissue-derived stem/stromal cell infusion [time frame: 24 months]

Thirty patients with recent-onset type 1 diabetes will receive adipose tissue-derived stem/stromal cells (ASCs). The patients will be admitted on the day of the infusion and will be discharged 24 h after infusion. A single dose of ASCs will be infused through the peripheral upper arm vein in 15–20 min.

Frequency of hypoglycemia (%) insulin dose/kg, and blood samples will be drawn for the glycated hemoglobin assessment (high-performance liquid chromatography by boronate affinity) at each follow-up outpatient visit (1, 3, 6, 12, 18, and 24 months after the ASC infusion)

Secondary Outcome Measures:
1. Oral cholecalciferol 2000 UI/day supplementation [time frame: 24 months]

The same patients with recent-onset type 1 diabetes that will receive adipose tissue-derived stem/stromal cells (ASCs) will also receive oral cholecalciferol 2000 UI/day for 24 months.

Serum 25-hydroxy vitamin D will be assessed at each follow-up outpatient visit (1, 3, 6, 12, 18, and 24 months after the ASC infusion) and C-peptide will be analyzed by immunofluorometric assay, considering the time 0 (basal), and peak stimulated C-peptide (30, 60, 90, and 120 min) after the mixed meal test (Glucerna).

Ages Eligible for Study:	16 Years to 35 Years (Child, Adult)
Sexes Eligible for Study:	All
Accepts Healthy Volunteers:	No

Criteria
Inclusion Criteria:
- Diagnosis of type 1 diabetes according to the American Diabetes Association criteria for a period of less than 4 months
- Pancreatic autoimmunity (positive anti-glutamic acid decarboxylase [GAD] and/or islet antigen 2 [anti-IA2])

Exclusion Criteria:
- Clinical evidence of malignancy or prior history
- Pregnancy or desire to become pregnant within 12 months of the study
- Breastfeeding
- HIV(+), hepatitis B (+), hepatitis C(+)
- Diabetic ketoacidosis at diagnosis
- Glomerular filtration rate of less than 60 mL/min
- Use of immunosuppressors or glucocorticoids

Publications:
Araujo DB, Dantas JR, Silva KR, Souto DL, Pereira MFC, Moreira JP, Luiz RR, Claudio-Da-Silva CS, Gabbay MAL, Dib SA, Couri CEB, Maiolino A, Rebelatto CLK, Daga DR, Senegaglia AC, Brofman PRS, Baptista LS, Oliveira JEP, Zajdenverg L, Rodacki M. Allogenic adipose tissue-derived stromal/stem cells and vitamin D supplementation in patients with recent-onset type 1 diabetes mellitus: a 3-month follow-up pilot study. Front Immunol. 2020;11:993. https://doi.org/10.3389/fimmu.2020.00993. eCollection 2020.

Part IX
Diabetes Mellitus: Type 1 Diabetes (Canada)

Chapter 25
Stem Cell Mobilization (Plerixafor) and Immunologic Reset in Type 1 Diabetes (T1DM)

Brief Summary:

Type 1 diabetes is an autoimmune disease characterized by destruction of pancreatic beta-cells, resulting in the absolute deficiency of insulin. Presently, there is no known cure for this disease.

Our proposed interventional trial is based on the "immunological reset" approach: T-depletion therapy and anti-inflammatory treatment will restore self-tolerance in T1DM patients; autologous, peripheral-blood mobilized hematopoietic CD34+-enriched stem cells, and a long-acting GLP-1 analogue will promote pancreatic islet regeneration and repair.

The short-term goals of this protocol is to demonstrate that subjects with new-onset T1DM undergoing autologous hematopoietic stem cell mobilization and immunologic reset will have greater preservation of endogenous insulin secretion compared with controls. Another objective is to investigate whether this nonmyeloablative treatment is safe, without the need for chronic immune suppression.

Condition or disease	Intervention/treatment	Phase
Diabetes mellitus, type 1	Drug: Plerixafor Drug: Alemtuzumab Drug: Anakinra Drug: Etanercept Drug: Liraglutide	Phase 1 Phase 2

University of Alberta, Edmonton, Alberta, Canada
ClinicalTrials.gov Identifier: NCT03182426

© The Author(s), under exclusive license to Springer Nature
Switzerland AG 2022
J. N. Weiss, *Stem Cell Surgery Trials in Heart Failure and Diabetes*,
https://doi.org/10.1007/978-3-030-78010-4_25

Study Type:	Interventional (Clinical Trial)
Estimated Enrollment:	60 Participants
Allocation:	Randomized
Intervention Model:	Parallel Assignment
Actual Study Start Date:	August 15, 2017
Estimated Primary Completion Date:	August 30, 2022
Estimated Study Completion Date:	December 31, 2022

Arm	Intervention/treatment
Experimental: Treated arm For participants assigned to the treated arm, they will follow the study regime below: Intervention treatment will last from Day 0 up to Month 24. Day 0: Subjects will receive alemtuzumab (30 mg iv single dose), anakinra (100 mg sc), etanercept (50 mg sc), and liraglutide (0.6 mg sc) Day 1: Subjects will receive plerixafor (0.24 mg/kg/day) sc to mobilize CD34+ stem cells to the peripheral blood Day 1: Continuation with anakinra 100 mg sc daily for 12 months; etanercept 50 mg sc twice weekly for the first 3 months and 50 mg sc weekly for another 9 months; liraglutide 0.6 mg sc daily for 7 days and then 1.2 mg sc daily (or up to 1.8 mg daily) as tolerated for 24 months	Drug: Plerixafor Systemic CD34+ stem cell mobilization for beta-cell repair Other name: Mozobil Drug: Alemtuzumab T-cell depletion Other name: Lemtrada Drug: Anakinra Anti-inflammatory Other name: Kineret Drug: Etanercept Anti-inflammatory Other name: Enbrel Drug: Liraglutide Beta-cell regeneration Other name: Victoza
Experimental: Control arm For participant assigned to the control arm, they will be monitored and tested for the first 12 months and receive intervention treatment from Month 12 up to Month 24 Month 12: Subjects will receive alemtuzumab (30 mg iv single dose), anakinra (100 mg sc), etanercept (50 mg sc), and liraglutide (0.6 mg sc) Month 12 + 1 day: Subjects will receive plerixafor (0.24 mg/kg/day) sc Month 12 + 1 day: Continuation with anakinra 100 mg sc daily for 12 months; etanercept 50 mg sc twice weekly for the first 3 months, and 50 mg sc weekly for another 9 months; liraglutide 0.6 mg sc daily for 7 days and then 1.2 mg sc daily (or up to 1.8 mg daily) as tolerated for 12 months	Drug: Plerixafor Systemic CD34+ stem cell mobilization for beta-cell repair Other name: Mozobil Drug: Alemtuzumab T-cell depletion Other name: Lemtrada Drug: Anakinra Anti-inflammatory Other name: Kineret Drug: Etanercept Anti-inflammatory Other name: Enbrel Drug: Liraglutide Beta-cell regeneration Other name: Victoza

Primary Outcome Measures:
1. Change of 2-h mixed meal-stimulated C-peptide AUC [time frame: baseline and Months 3, 6, 9, 12, 18, and 24]

This AUC will be normalized by dividing it by 120 min (the number of minutes over which it is determined) and will be adjusted through the inclusion of baseline C-peptide AUC as a covariate in the analysis.

2. Rate of serious adverse event/medical event of special interest [time frame: within 24 months]

Secondary Outcome Measures:
1. "Responder" status [time frame: Months 3, 6, 9, 12, 18, and 24]

A subject is considered a responder if at the given time point, the subject has (a) HbA1c $\leq$6.5% and (b) mean daily insulin use <0.5 IU/kg/day over 7 consecutive days during the 2 weeks preceding the visit.

2. Exogenous insulin usage [time frame: baseline and Months 3, 6, 9, 12, 18, and 24]

Mean total daily insulin dose assessed over 7 consecutive days during the 2 weeks preceding the clinical visits.

3. Proportion of subjects with HbA1c $\leq$6.5% [time frame: baseline and Months 3, 6, 9, 12, 18, and 24]
4. Proportion of subjects with HbA1c $\leq$7.0% [time frame: baseline and Month 3, 6, 9, 12, 18, and 24]
5. Proportion of subjects free from severe hypoglycemia [time frame: baseline and Months 3, 6, 9, 12, 18, and 24]

Proportion of subjects free from severe hypoglycemia reported frequency of hypoglycemia by Hypo Score and Lability Index and CGMS

6. Proportion of subjects progressing to complete beta cell loss [time frame: baseline and Months 3, 6, 9, 12, 18, and 24]

Proportion of subjects who become C-peptide-negative

7. Autoantibodies associated with T1DM [time frame: baseline and Month 24 or the study withdrawal visit]

Including GAD, ICA512, IA2A, ZnT8, and mIAA

8. T1DM T-cell autoreactivity [time frame: baseline and Months 3, 6, 9, 12, 18, and 24]
9. T-cell phenotyping [time frame: baseline and Months 3, 6, 9, 12, 18, and 24]

Ages Eligible for Study:	18 Years to 45 Years (Adult)
Sexes Eligible for Study:	All
Accepts Healthy Volunteers:	No

Criteria
Inclusion Criteria:
- Patient is aged 18–45
- To be eligible, participants must have:

  – A clinical diagnosis of type 1 diabetes using the diagnostic criteria of the CDA

– Residual β-cell function, defined by stimulated C-peptide >0.6 but ≤10.5 ng/mL on MMTT
– One or more positive autoantibodies (GAD, ICA512, IA2A, ZnT8, mIAA) to confirm T1DM
– No underlying condition that would preclude enrollment at the PI's discretion

- Participants must be capable of understanding the purpose and risks of the study and must sign a statement of informed consent, with additional parental consent where required.

Exclusion Criteria:
- Duration of T1DM longer than 180 days.
- Severe co-existing cardiac disease, characterized by any one of these conditions: (a) recent myocardial infarction (within the past 6 months); (b) left ventricular ejection fraction <30%; or (c) evidence of ischemia on functional cardiac exam.
- Active alcohol or substance abuse, including cigarette smoking (must be abstinent for 6 months prior to the study enrolment).
- Psychiatric disorder making the subject not a suitable candidate for this study (e.g., schizophrenia, bipolar disorder, or major depression that is unstable or uncontrolled on current medication).
- History of non-adherence to prescribed regimens.
- Hypersensitivity to any of the required study medications.
- Significant systemic infection during the 3 weeks before the start of the study intervention (e.g., infection requiring hospitalization, major surgery, or IV antibiotics to resolve; other infections, e.g., bronchitis, sinusitis, localized cellulitis, candidiasis, or urinary tract infections, must be assessed on a case-by-case basis by the investigator regarding whether they are serious enough to warrant exclusion).
- Active infection including hepatitis C, hepatitis B, HIV, tuberculosis (subjects with a positive PPD performed within 1 year of enrollment and no history of adequate chemoprophylaxis).
- Any history of current malignancies, other than non-melanoma skin cancer (to be included in the study, the subject must have had fewer than 5 occurrences of non-melanoma skin cancer, and the last occurrence must not be within 3 months of the study entry).
- BMI >35 kg/m^2 at the screening visit.
- Age of less than 18 or greater than 45 years.
- Measured glomerular filtration rate (GFR) <60 m/min/1.73 m^2.
- Presence or history of macroalbuminuria (>300 mg/g creatinine).
- Clinical suspicion of nephritic (hematuria, active urinary sediment) or rapidly progressing renal impairment (e.g., increase in serum creatinine of 25% within the last 3–6 months).
- Baseline Hb <105 g/L in women or <120 g/L in men.
- Baseline screening liver function tests outside of the normal range, with the exception of uncomplicated Gilbert's syndrome. An initial LFT panel with any

values >1.5 times the upper limit of normal (ULN) will exclude a patient without a re-test; a re-test for any values between ULN and 1.5 times ULN should be made, and if the values remain elevated above the normal limits, the patient will be excluded.

- Untreated proliferative retinopathy.
- Positive pregnancy test, intent for future pregnancy or male subjects' intent to procreate, failure to follow effective contraceptive measures, or presently breastfeeding.
- EBV viral load of >10,000 copies per 106 peripheral blood mononuclear cells (PBMCs) as determined by quantitative polymerase chain reaction (qPCR). If there is any clinical suspicion that a subject who is EBV-seronegative and with EBV PCR <10,000 copies per 106 PBMCs has symptoms consistent with infectious mononucleosis prior to administration of study treatment, then a monospot test result must be negative before the subject can be enrolled.
- Positive result on the rapid plasma reagin (RPR) test for syphilis; if the result of the RPR test is positive, a negative confirmatory test, is required (e.g., fluorescent treponemal antibody absorbed [FTA-ABS] test).
- History of using any investigational drug within the 3 months before enrollment in this study.
- History of using any potent immunosuppressive agent (e.g., systemic high-dose corticosteroids on a chronic basis, methotrexate, cyclosporine, or anti-TNF agents) within the 30 days before the study treatment.
- History of receiving any live vaccine within the 30 days before the study treatment.
- Any major surgical procedure within 30 days before the study treatment.
- Insulin requirement >1.0 U/kg/day.
- HbA1C >12% at screening.
- Uncontrolled hyperlipidemia [fasting LDL cholesterol >3.4 mmol/L, treated or untreated; and/or fasting triglycerides >2.3 mmol/L].
- Under treatment for a medical condition requiring chronic use of steroids.
- Use of coumadin or other anticoagulant therapy (except aspirin) or subject with PT INR >1.5.
- Untreated celiac disease.
- Patients with Graves' disease unless previously adequately treated with radioiodine ablative therapy
- Family or personal history of multiple endocrine neoplasia type 2 or medullary thyroid carcinoma
- Hypersensitive to *E. coli*-derived protein.
- Clinically significant abnormal lab values during the screening period, other than those due to T1DM. Permitted ranges for selected lab values are presented in the table below. A clinically significant abnormal value will not result in exclusion if, upon re-test, the abnormality is resolved or becomes clinically insignificant.

Part X
Diabetes Mellitus: Type 1 Diabetes (Middle East)

Chapter 26
Mesenchymal Stem Cell Transplantation in Newly Diagnosed Type 1 Diabetes Patients (MSCTXT1DM)

Brief Summary:

Study Objects: Diabetes is an autoimmune disease that is mainly caused by an immune reaction to beta cells in the pancreas. In this study, mesenchymal stem cells will be used for immune response modulation and regeneration improvement.

Study Design and Method: In a triple-blinded, randomized, placebo-controlled phase I/II clinical trial, 20 patients with newly diagnosed type 1 diabetes, who would be visited in the Children's Growth and Development Research Center of Tehran University of Medical Sciences and Royan Institute Cell Therapy Center, would be assessed through two groups, the case group and the placebo group.

Participants: The participants include patients of both sexes aged 8–40 years old who are diagnosed with type 1 diabetes not more than 6 weeks, have antibody against beta cells diagnosed in their blood, have fasting C-peptide more than or equal to 0.3 ng/mL, and are not suffering from other acute or chronic diseases and cancers.

Interventions: Intravascular transplantation of autologous mesenchymal stem cells in the case group and placebo injection in the control group

Outcome variables: Safety and efficacy

Condition or disease	Intervention/treatment	Phase
Diabetes mellitus Diabetes mellitus, type 1 Diabetes mellitus, insulin-dependent	Biological: Intravenous injection of autologous mesenchymal stem cells Other: Intravenous injection of placebo	Phase 1 Phase 2

Detailed Description:

Diabetes is an autoimmune disease that is mainly caused by an immune reaction to beta cells in the pancreas. At present, insulin injection is a routine treatment for

Royan Institute Tehran, Islamic Republic of Iran
ClinicalTrials.gov Identifier: NCT04078308

J. N. Weiss, *Stem Cell Surgery Trials in Heart Failure and Diabetes*, https://doi.org/10.1007/978-3-030-78010-4_26

diabetes. Although injected insulin maintains blood glucose, this method cannot result in physiologic reaction to blood glucose changes. Moreover, patients experience diabetes complications such as neuropathy, nephropathy, visual and cardiovascular problems, and hypoglycemic unawareness. Therefore, based on previous studies, a treatment option that leads to pancreatic beta cell restoration and inhibits the immune response to these cells could be a hopeful clinical choice.

In this clinical trial, autologous bone marrow-derived mesenchymal stem cells will be used for immune response modulation and regeneration improvement. Hence, based on the inclusion and exclusion criteria, 20 patients with type 1 diabetes will be selected. After clarifying the procedure and fulfilling the agreement to participate in this trial, they will be divided into two groups.

Bone marrow is aspirated from the patient's bone, and after isolation of mesenchymal stem cells and characterization of these cells, patients in the case group will be intravenously injected with 1 million autologous mesenchymal stem cells per kilogram of the patient's body weight in each dose in weeks 0 and 3, whereas the control group will receive a placebo. Then the patients will be followed up for 1 year. During this time, different parameters would be evaluated in weeks 1, 2, and 4 and Months 2, 3, 6, 9, and 12. Laboratory screenings will be done during this period to evaluate the safety and efficacy of this treatment.

Study Type:	Interventional (Clinical Trial)
Estimated Enrollment:	20 Participants
Allocation:	Randomized
Intervention Model:	Crossover Assignment

Actual Study Start Date:	July 6, 2015
Estimated Primary Completion Date:	September 26, 2019
Estimated Study Completion Date:	April 1, 2020

Arm	Intervention/treatment
Experimental: Mesenchymal stem cell transplantation Patients with type 1 diabetes who will receive intravenous injection of autologous bone marrow-derived mesenchymal stem cells	Biological: Intravenous injection of autologous mesenchymal stem cells Intravenous injection of 1 million bone marrow-derived autologous mesenchymal stem cells (MSCs) per kilogram of the patient's body weight in each dose in weeks 0 and 3
Placebo comparator: Placebo Patients with type 1 diabetes who will receive intravenous injection of normal saline (sodium chloride 0.9%)	Other: Intravenous injection of placebo Intravenous injection of normal saline (sodium chloride 0.9%)

Primary Outcome Measures:
1. Number of participants with treatment-related adverse events as assessed by CTCAE v5.0 [time frame: 12 months after the first infusion]

Safety will be assessed by evaluating patients based on CTCAE (v5.0) to assess treatment-related adverse events after intravenous transplantation of autologous bone marrow-derived mesenchymal stem cells

2. Change from the baseline number of hypoglycemic unawareness episodes at 12 months after intravenous transplantation of autologous bone marrow-derived mesenchymal stem cells [time frame: 12 months after the first infusion]

Number of hypoglycemic unawareness episodes will be assessed by evaluating patients' blood glucose monitoring sheets

Secondary Outcome Measures:
1. Change from baseline fasting blood sugar (FBS) at 12 months after intravenous transplantation of autologous bone marrow-derived mesenchymal stem cells [time frame: 12 months after the first infusion]

Assessing fasting blood sugar (FBS) at 12 months after intravenous transplantation of autologous bone marrow-derived mesenchymal stem cells and comparing the results with baseline fasting blood sugar (FBS)

2. Change from baseline C-peptide at 12 months after intravenous transplantation of autologous bone marrow-derived mesenchymal stem cells [time frame: 12 months after the first infusion]

Assessing serum C-peptide at 12 months after intravenous transplantation of autologous bone marrow-derived mesenchymal stem cells and comparing the results with baseline serum C-peptide level

3. Change from baseline HbA1C at 12 months after intravenous transplantation of autologous bone marrow-derived mesenchymal stem cells [time frame: 12 months after the first infusion]

Assessing HbA1C at 12 months after intravenous transplantation of autologous bone marrow-derived mesenchymal stem cells and comparing the results with baseline HbA1C

4. Change from baseline 2-h postprandial blood glucose at 12 months after intravenous transplantation of autologous bone marrow-derived mesenchymal stem cells [time frame: 12 months after the first infusion]

Assessing 2-h postprandial blood glucose at 12 months after intravenous transplantation of autologous bone marrow-derived mesenchymal stem cells and comparing the results with baseline 2-h postprandial blood glucose

5. Change from baseline daily dose of exogenous insulin injected by patients (IU/kg/day) at 12 months after intravenous transplantation of autologous bone marrow-derived mesenchymal stem cells [time frame: 12 months after the first infusion]

Exogenous insulin requirement of patients will be measured based on their daily insulin injection report sheets; the total injected insulin units per day will be divided by the patient's weight in order to be comparable between the different patients.

6. Change from baseline lability index (LI) at 12 months after intravenous transplantation of autologous bone marrow-derived mesenchymal stem cells [time frame: 12 months after the first infusion]

Assessing 2 weeks of blood glucose report sheets, calculation is based on the changes in blood glucose levels over time.

7. Change from the baseline SF-36 Quality of life (QOL) questionnaire score at 12 months after intravenous transplantation of autologous bone marrow-derived mesenchymal stem cells [time frame: 12 months after the first infusion]

Assessing change in patients' quality of life by answering the SF-36 questionnaire before transplantation and 12 months after intravenous transplantation of autologous bone marrow-derived mesenchymal stem cells. This questionnaire asks about the general aspects of the patients' life.

The results would be reported as the total score, and the scale range is from 0% to 100%. 0% is considered as the worst condition and 100% as the best condition.

8. Change from the baseline Diabetes-Specific Quality of life (DQOL) questionnaire score at 12 months after intravenous transplantation of autologous bone marrow-derived mesenchymal stem cells [time frame: 12 months after the first infusion]

Assessing change in patients' quality of life by answering the Diabetes-Specific Quality of Life (DQOL) questionnaire before transplantation and 12 months after intravenous transplantation of autologous bone marrow-derived mesenchymal stem cells. This questionnaire asks about different aspects of patients' life in relation to diabetes.

The results would be reported as the total score, and the scale range is from 0% to 100%. 0% is considered as the worst condition and 100% as the best condition.

9. Changes from baseline autoantibody levels in patients' blood at 12 months after intravenous transplantation of autologous bone marrow-derived mesenchymal stem cells [time frame: 12 months after the first infusion]

Assessing patients' islet cell antibody (ICA), glutamic acid decarboxylase antibody (GADA), and insulinoma-associated protein-2 antibody (IA-2A) levels after intravenous transplantation of autologous bone marrow-derived mesenchymal stem cells

10. Changes from baseline serum cytokines levels in patients' blood at 12 months after intravenous transplantation of autologous bone marrow-derived mesenchymal stem cells [time frame: 12 months after the first infusion]

Assessing the serum cytokine levels of patients after intravenous transplantation of autologous bone marrow-derived mesenchymal stem cells

Ages Eligible for Study:	8 Years to 40 Years (Child, Adult)
Sexes Eligible for Study:	All
Accepts Healthy Volunteers:	No

Criteria

Inclusion Criteria:
- Type 1 diabetes detection in less than 6 weeks
- Diabetes diagnosis according to the American Diabetes Association (ADA)
- Presence of antibodies against pancreatic beta cells
- Fasting C-peptide $\geq$0.3 ng/mL

Exclusion Criteria:
- Pregnancy or breastfeeding
- Cancer
- Any acute or severe disease (according to the physicians' diagnosis, such as cardiac, pulmonary, hepatic, kidney, mental, ... diseases)
- Positive results for human immunodeficiency virus (HIV), human T-lymphotropic virus (HTLV), hepatitis B virus (HBV), hepatitis C virus (HCV), cytomegalovirus (CMV)
- Immune-deficient or hyperesthesia
- History of severe ketoacidosis

Chapter 27
Use of Stem Cells in Diabetes Mellitus Type 1

Brief Summary:

Allogeneic adipose-derived mesenchymal stem cells will be injected into patients newly diagnosed with type 1 diabetes mellitus

Condition or disease	Intervention/treatment	Phase
Diabetes mellitus type 1	Biological: Adipose mesenchymal cells with bone marrow mononuclear cells	Phase 1

Detailed Description:

Adipose-derived stem cells (ASCs) are to be collected from blood group O donor cells will be passaged to passage 5. Before the samples are released, they will be subject to testing for any bacterial or fungal growth as well as endotoxin and mycoplasma. All these tests must be negative. In addition, the cell count and viability (must be more than 80%) are done before release. Surface marker documentation is done on the cells before the release via flow cytometry.

The cells should be infused within 2 h of release. The dose is to be repeated after 6 months of a total of two doses in patients. The tests and follow-up visits are to be done at weeks 12, 24, and 36 when the study is stopped. These cells will be injected intravenously into patients newly diagnosed with type 1 diabetes mellitus.

Study Type:	Interventional (Clinical Trial)
Estimated Enrollment:	20 participantsParticipants
Allocation:	Non-randomized
Intervention Model:	Parallel Assignment

Cell Therapy Center Amman, Jordan
ClinicalTrials.gov Identifier: NCT02940418

J. N. Weiss, *Stem Cell Surgery Trials in Heart Failure and Diabetes*, https://doi.org/10.1007/978-3-030-78010-4_27

Actual Study Start Date:	February 19, 2017
Estimated Primary Completion Date:	November 2019
Estimated Study Completion Date:	January 2020

Arm	Intervention/treatment
Active comparator: Dose 1 mil/kg Adipose mesenchymal cells with bone marrow mononuclear cells Two doses with a 6-month interval of allogeneic adipose mesenchymal cells +1 mil/kg of autologous bone marrow mononuclear cells will be injected intravenously	Biological: Adipose mesenchymal cells with bone marrow mononuclear cells Allogeneic adipose-derived mesenchymal cells with autologous bone marrow mononuclear cells will be injected
Active comparator: Dose 10 mil/kg Adipose mesenchymal cells with bone marrow mononuclear cells Two doses with a 6-month interval of allogeneic adipose mesenchymal cells +10 mil/kg of autologous bone marrow mononuclear cells will be injected intravenously	Biological: Adipose mesenchymal cells with bone marrow mononuclear cells Allogeneic adipose-derived mesenchymal cells with autologous bone marrow mononuclear cells will be injected

Primary Outcome Measures:

1. Safety of using allogeneic ASC assessed by any adverse events [time frame: 6 months]

   Patients will be assessed for any adverse events as a result of the injection.

Ages Eligible for Study:	18 Years to 35 Years (Adult)
Sexes Eligible for Study:	All
Accepts Healthy Volunteers:	No

Criteria

Inclusion Criteria:

1. Adult patients with type 1 diabetes mellitus
2. Patients aged 18–35 years, both sexes
3. Duration of disease: not exceeding 3 years unless C-peptide is not less than 0.5 ng/mL
4. C-peptide at inclusion baseline should not be less than 0.5 ng/mL
5. No clinical evidence of renal, retinal, vascular, or skin complications
6. Body mass index not exceeding 30
7. Any HbA1c
8. At least one positive antibody, either anti-glutamic acid decarboxylase-65 or insulinoma-associated-2 autoantibodies (anti-1A2)
9. Informed consent by patient

Exclusion Criteria:
1. Patients aged less than 18 years and more than 35 years
2. Pregnancy
3. Married women or women expected to be married within the study period
4. History of allergy, cancer, bronchial asthma, liver disease, or hepatitis
5. Diabetic coma or pre-coma current or recent within the last 2 months
6. C-peptide less than 0.5 ng/mL
7. Disease duration of more than 3 years
8. Complication mentioned in 5 above in inclusion
9. Non-consenting patient or withdrawal of consent
10. Bleeding disorders

Part XI
Diabetes Mellitus: Type 1 Diabetes (Sweden)

Chapter 28
Wharton's Jelly-Derived Mesenchymal Stromal Cell-Repeated Treatment of Adult Patients Diagnosed with Type I Diabetes

Brief Summary:

An open-label, parallel, single-centered trial of Wharton's Jelly derived allogeneic mesenchymal stromal cell-repeated treatment to preserve endogenous insulin production in adult patients diagnosed with type 1 diabetes

Condition or disease	Intervention/treatment	Phase
Type1 diabetes	Drug: ProTrans	Phase 1 Phase 2

Study Type:	Interventional (Clinical Trial)
Estimated Enrollment:	18 Participants
Allocation:	N/A
Intervention Model:	Single-Group Assignment

Actual Study Start Date:	May 17, 2019
Estimated Primary Completion Date:	October 17, 2020
Estimated Study Completion Date:	October 17, 2024

Arm	Intervention/treatment
Experimental: ProTrans-Repeat Patients 1–3: 25×10^6 cells Patients 4–6: 100×10^6 cells Patients 7–9: 200×10^6 cells Control group 9 patients = non-treated	Drug: ProTrans ProTrans is a pooled allogeneic mesenchymal stromal cell treatment

Karolinska Trial Alliance, Fas 1 enheten, Karolinska Universitetssjukhuset Huddinge, Huddinge, Sweden
ClinicalTrials.gov Identifier: NCT03973827

Primary Outcome Measures:
1. The primary safety endpoint in this study is; safety parameters include adverse events, hypoglycemia and allergic reactions [time frame: 372 days]

To investigate the safety and tolerance after a repeated allogeneic infusion of Wharton's Jelly mesenchymal stem cells (WJMSCs) intravenously in adult patients diagnosed with type 1 diabetes after 1 year following the repeated treatment.

2. Delta change of C-peptide area under the curve (AUC) (0–120 min) for mixed meal tolerance test (MMTT) at day 372 following WJMSC infusion when compared with the test performed before the start of treatment [time frame: 372 days]

Secondary Outcome Measures:
1. Number of patients independent on insulin (ADA criteria) at day 372 [time frame: 372 days]
2. Number of patients with daily insulin needs <0.25 U/kg at day 372
3. [time frame: 372 days]
4. HbA1c at day 372 [time frame: 372 days]
5. Glucose variability (mean amplitude of glycemic excursions and glycemic lability index) duration derived from the continuous glucose monitoring system® at day 372 [time frame: 372 days]
6. Delta change of the levels of fasting C-peptide at day 372 when compared with the test before the start of treatment [time frame: 372 days]
7. Numbers of patients with peak C-peptide >0.20 nmol/L, in response to the MMTT, at day 372 [time frame: 372 days]

Ages Eligible for Study:	18 Years to 64 Years (Adult)
Sexes Eligible for Study:	Male
Accepts Healthy Volunteers:	No

Criteria
Inclusion Criteria:
1. A new written informed consent for participation of the study is required to be given before undergoing any study-specific procedures.
2. Only patients that have previously been dosed by the IMP according to protocol Protrans-1 are eligible for a second dose of Protrans.
3. No identified IMP related ongoing adverse event, neither history of any adverse event that is evaluated potentially to be related to the previous IMP dosing in Protrans I.
4. Clinical history compatible with type 1 diabetes diagnosed less than 3 years before enrolment. This also includes control patients not receiving IMP.
5. Only male patients aged between 18 and 41 years will be included.
6. Mentally stable and, in the opinion of the investigator, able to comply with the procedures of the study protocol.

Exclusion Criteria:

1. Inability to provide informed consent
2. Patients with body mass index (BMI) >30 or weight >100 kg
3. Patients with weight <50 kg
4. Patients with unstable cardiovascular status including NYHA class III/IV or symptoms of angina pectoris
5. Patients with uncontrolled hypertension (≥160/105 mmHg)
6. Patients with active infections unless treatment is not judged necessary by the investigators
7. Patients with latent or previous as well as ongoing therapy against tuberculosis or exposed to tuberculosis or has travelled in areas with a high risk of tuberculosis or mycosis within the last 3 months
8. Patients with serological evidence of infection with HIV, *Treponema pallidum*, hepatitis B antigen (patients with serology consistent with previous vaccination and a history of vaccination are acceptable) or hepatitis C
9. Patients with any immunosuppressive treatment
10. Patients with known demyelinating disease or with symptoms or physical examination findings consistent with possible demyelinating disease
11. Patients with known, or previous, malignancy
12. Taking oral anti-diabetic therapies or any other concomitant medication that may interfere with glucose regulation other than insulin
13. Patients with GFR <80 mL/min/1.73 m² body surface
14. Patients with proliferative retinopathy
15. Patient with any condition or any circumstance that in the opinion of the investigator would make it unsafe to undergo treatment with MSC
16. Known hypersensitivity against any excipients, i.e., dimethyl sulfoxide (DMSO)

Part XII
Diabetes Mellitus: Type 2 Diabetes (United States)

Chapter 29
Allogeneic Mesenchymal Human Stem Cell Infusion Therapy for Endothelial Dysfunction in Diabetic Subjects (ACESO)

Brief Summary:
This is a 16-subject trial to demonstrate the safety of allogeneic hMSCs administered via infusion therapy for diabetic subjects with endothelial dysfunction.

Condition or disease	Intervention/treatment	Phase
Diabetes mellitus	Drug: 20 million allogeneic mesenchymal human stem cells	Phase 1
	Drug: 100 million allogeneic mesenchymal human stem cells	Phase 2

Detailed Description:
Sixteen (16) diabetic subjects with endothelial dysfunction will be scheduled to undergo a peripheral intravenous infusion after meeting all the inclusion/exclusion criteria at baseline.

Eight (8) subjects will be treated with 20 million (2×10^7) allogeneic hMSCs, and eight (8) subjects will be treated with 100 million (100×10^6) allogeneic hMSCs.

Follow-up: Subjects will be followed at 3, 7, 14, 28 days, and 3 months post-infusion to complete all the safety and efficacy assessments and at 6 and 12 months post-infusion to complete the safety assessments.

Study Type:	Interventional (Clinical Trial)
Estimated Enrollment:	10 Participants
Allocation:	Randomized
Intervention Model:	Parallel Assignment

University of Miami Miller School of Medicine, Miami, Florida, USA
ClinicalTrials.gov Identifier: NCT02886884

Actual Study Start Date:	October 20, 2017
Estimated Primary Completion Date:	September 2023
Estimated Study Completion Date:	September 2024

Arms and Interventions

Arm	Intervention/treatment

Experimental: 20 million allogeneic hMSCs
Eight (8) subjects will be administered 20 million allogeneic mesenchymal human stem cells (hMSCs) via peripheral intravenous infusion
	Drug: 20 million allogeneic mesenchymal human stem cells
	1 single intravenous infusion
	Other names:
	    Allo-hMSCs
	    Stem cells

Experimental: 100 million hMSCs
Eight (8) subjects will be administered 100 million allogeneic mesenchymal human stem cells (hMSCs) via peripheral intravenous infusion
	Drug: 100 million allogeneic mesenchymal human stem cells
	1 single intravenous infusion
	Other names:
	    Allo-hMSCs
	    Stem cells

Primary Outcome Measures:
1. Assess treatment-emergent serious adverse events [time frame: 1 month post-infusion]

Defined as the composite of death, non-fatal pulmonary embolism, stroke, hospitalization for worsening dyspnea, and clinically significant laboratory test abnormalities, determined per the investigator's judgment

Secondary Outcome Measures:
1. Assess endothelial progenitor cell colony-forming units [time frame: assess at baseline; Days 3, 7, and 14; Day 28; and Month 3 following IV allogeneic mesenchymal stem cell infusion]

Assess endothelial progenitor cell colony-forming units

2. Assess circulating inflammatory markers (IL-1) [time frame: assess at baseline; Days 3, 7, and 14; Day 28; and Month 3 following IV allogeneic mesenchymal stem cell infusion]

Assess circulating inflammatory markers (IL-1)

3. Assess circulating inflammatory markers (IL-6) [time frame: assess at baseline; Days 3, 7, and 14; Day 28; and Month 3 following IV allogeneic mesenchymal stem cell infusion]

Assess circulating inflammatory markers (IL-6)

4. Assess circulating inflammatory markers (tumor necrosis factor (TNF) alpha) [time frame: assess at baseline; Days 3, 7, and 14; Day 28; and Month 3 following IV allogeneic mesenchymal stem cell infusion]

Assess circulating inflammatory markers (tumor necrosis factor (TNF) alpha)

5. Assess circulating inflammatory marker (CRP) [time frame: assess at baseline; Days 3, 7, and 14; Day 28; and Month 3 following IV allogeneic mesenchymal stem cell infusion]

Assess circulating inflammatory markers (CRP)

6. Assess circulating angiogenic factors [time frame: assess at baseline; Days 3, 7, and 14; Day 28; and Month 3 following IV allogeneic mesenchymal stem cell infusion]

Assess circulating angiogenic factors known to mobilize and recruit endothelial progenitor cell (EPC) (VEGF, stromal cell-derived factor (SDF)-1 alpha, and SCF)

7. Assess stromal cell-derived factor (SDF)-1 alpha [time frame: assess at baseline; Days 3, 7, and 14; Day 28; and Month 3 following IV allogeneic mesenchymal stem cell infusion]

Assess circulating angiogenic factors known to mobilize and recruit endothelial progenitor cell (EPC) such as stromal cell-derived factor (SDF)-1 alpha

8. Assess vascular endothelial growth factor (VEGF) [time frame: assess at baseline; Days 3, 7, and 14; Day 28; and Month 3 following IV allogeneic mesenchymal stem cell infusion]

Assess circulating angiogenic factors known to mobilize and recruit endothelial progenitor cell (EPC) such as vascular endothelial growth factor (VEGF)

9. Assess stem cell factor (SCF) [time frame: assess at baseline; Days 3, 7, and 14; Day 28; and Month 3 following IV allogeneic mesenchymal stem cell infusion]

Assess circulating angiogenic factors known to mobilize and recruit endothelial progenitor cell (EPC) such as stem cell factor (SCF)

10. Assess FMD% [time frame: assess at baseline; Days 3, 7, and 14; Day 28; and Month 3 following IV allogeneic mesenchymal stem cell infusion]

Assess FMD% change

Ages Eligible for Study:	21 Years to 90 Years (Adult, Older Adult)
Sexes Eligible for Study:	All
Accepts Healthy Volunteers:	No

Criteria

Inclusion Criteria:

- Age of ≥ 21 and <90 (inclusive) years.
- Written informed consent.
- Have endothelial dysfunction defined by impaired flow-mediated vasodilation (FMD $<7\%$).
- Have an ejection fraction $>45\%$ by gated blood pool scan, two-dimensional echocardiogram, cardiac MRI, cardiac CT, or left ventriculogram within the previous 3 months.
- Have diabetes mellitus type 2 documented by hemoglobin adult type 1 component (A1C) $>7\%$ or being on medical therapy for diabetes.
- Females of childbearing potential must use two forms of birth control for the duration of the study. Female subjects must undergo a blood or urine pregnancy test at screening and within 36 h prior to infusion.

Exclusion Criteria:

- In order to participate in this study, a subject must not:
- Be younger than 21 years or older than 90 years of age
- Have a baseline glomerular filtration rate <35 mL/min 1.73 m^2 estimated using the Modification of Diet in Renal Disease (MDRD) formula
- Have poorly controlled blood glucose levels with hemoglobin A1C $>8.5\%$.
- Have a history of proliferative retinopathy or severe neuropathy requiring medical treatment
- Have a hematologic abnormality as evidenced by hematocrit $<25\%$, white blood cell <2500/uL, or platelet values $<100,000$/uL without another explanation
- Have liver dysfunction, as evidenced by enzymes (AST and ALT) greater than three times the upper limit of normal
- Have a bleeding diathesis or coagulopathy (INR >1.3), cannot be withdrawn from anticoagulation therapy, or will refuse blood transfusions
- Have lymphadenectomy or lymph node dissection in the right arm
- Be an organ transplant recipient or have a history of organ or cell transplant rejection.
- Have a clinical history of malignancy within the past 5 years (i.e., subjects with prior malignancy must be disease-free for 5 years), except curatively – treated basal cell or squamous cell carcinoma or cervical carcinoma
- Have a condition that limits lifespan to <1 year
- Have a history of drug or alcohol abuse within the past 24 months
- Be on chronic therapy with immunosuppressant medication, such as corticosteroids or tumor necrosis factor-alpha (TNFα) antagonists

- Be serum-positive for HIV, syphilis-VDRL (confirmation with FTA-ABS if needed (syphilis)), hepatitis B surface antigen, or viremic hepatitis C
- Be currently participating (or participated within the previous 30 days) in an investigational therapeutic or device trial
- Be pregnant, nursing, or of childbearing potential while not practicing effective contraceptive methods
- Have any other condition that in the judgment of the investigator would be a contraindication to enrollment or follow-up

Chapter 30
Patient-Derived Stem Cell Therapy for Diabetic Kidney Disease

Brief Summary:

The researchers will assess the safety, tolerability, dosing effect, and early signals of efficacy of intra-arterially delivered autologous (from self) adipose (fat) tissue-derived mesenchymal stem/stromal cells (MSC) in patients with progressive diabetic kidney disease (DKD).

Condition or disease	Intervention/treatment	Phase
Diabetic kidney disease Diabetic nephropathies Diabetes mellitus, type 2 Diabetes mellitus, type 1 Chronic kidney disease Diabetic nephropathy type 2 Kidney failure Kidney insufficiency	Biological: Autologous adipose-derived mesenchymal stem/stromal cells (MSC) Lower dose Biological: Autologous adipose-derived mesenchymal stem/stromal cells (MSC) Higher dose	Phase 1

Detailed Description:

This is a single-center, open-label, dose-escalating study assessing the safety, tolerability, dosing effect, and early signals of efficacy of intra-arterially delivered autologous (from self) adipose tissue-derived mesenchymal stem/stromal cells (MSC) in 30 patients with progressive diabetic kidney disease (DKD). DKD will be defined as chronic kidney disease (CKD; estimated glomerular filtration rate (eGFR) <60 mL/min/1.73 m^2) in the setting of diabetes mellitus (type 2; on anti-diabetes therapy) without overt etiologies of CKD beyond concomitant hypertension. Progressive DKD will be considered as eGFR 25–55 mL/min/1.73 m^2 with a) eGFR decline of 5 mL/min over 18 months or 10 mL/min over 3 years or b) an

Mayo Clinic in Rochester, Rochester, Minnesota, USA
ClinicalTrials.gov Identifier: NCT03840343

© The Author(s), under exclusive license to Springer Nature Switzerland AG 2022
J. N. Weiss, *Stem Cell Surgery Trials in Heart Failure and Diabetes*,
https://doi.org/10.1007/978-3-030-78010-4_30

157

intermediate or high 5-year risk of progression to end-stage kidney failure (dialysis or transplant) based on the validated Tangri 4-variable (age, sex, eGFR, urinary albumin-creatinine ratio) kidney failure risk equation.

Fifteen subjects will be placed in one of two cell dosage arms in a parallel design with single-kidney MSC administration at Day 0 and Month 3. Subjects will be followed up for a total of 15 months from the time of initial cell administration.

Study Type:	Interventional (Clinical Trial)
Estimated Enrollment:	30 Participants
Allocation:	Non-randomized
Intervention Model:	Parallel Assignment

Actual Study Start Date:	October 23, 2019
Estimated Primary Completion Date:	December 2024
Estimated Study Completion Date:	December 2025

Arm	Intervention/treatment
Experimental: Lower-dose MSC This arm will receive autologous adipose-derived mesenchymal stem/stromal cells (MSC) lower dose	Biological: Autologous adipose-derived mesenchymal stem/stromal cells (MSC) lower dose Two MSC infusions of 2.5×10^5 cells/kg at time zero and 3 months; single kidney, intra-arterial delivery
Experimental: Higher-dose MSC This arm will receive autologous adipose-derived mesenchymal stem/stromal cell (MSC) higher dose	Biological: Autologous adipose-derived mesenchymal stem/stromal cell (MSC) higher dose Two MSC infusions of 5.0×10^5 cells/kg at time zero and 3 months; single kidney, intra-arterial delivery

Primary Outcome Measures:

1. Adverse events [time frame: baseline through Month 15]

   The number of adverse events associated with MSC intervention per treatment arm

2. Adverse events [time frame: baseline through Month 15]

   The percentage of adverse events associated with MSC intervention per treatment arm

Secondary Outcome Measures:

1. Kidney function [time frame: baseline, Month 6]

   Change in measured glomerular filtration rate (mGFR). Measured as mL/min/BSA

2. Kidney function [time frame: pretreatment, Month 12]

   Change in the estimated glomerular filtration rate (eGFR) slope. Measured as mL/min/1.73 m²/month

Ages Eligible for Study:	45 Years to 75 Years (Adult, Older Adult)
Sexes Eligible for Study:	All
Accepts Healthy Volunteers:	No

Criteria

Inclusion Criteria:
- Diabetes mellitus (on anti-diabetes drug therapy)
- Age 45–75 years
- eGFR 25–55 mL/min/1.73 m² at the time of consent with (a) eGFR decline of 5 mL/min over 18 months or 10 mL/min over 3 years or (b) an intermediate or high 5-year risk of progression to end-stage kidney failure (dialysis or transplant) based on the validated Tangri 4-variable (age, sex, eGFR, urinary albumin-creatinine ratio) kidney failure risk equation (https://kidneyfailurerisk.com/)
- Primary cause of kidney disease is diabetes without suspicion of concomitant kidney disease beyond hypertension
- Spot urine albumin-creatinine ≥30 mg/g unless on RAAS inhibition
- Ability to provide informed consent

Exclusion Criteria:
- Hemoglobin A1c ≥11%
- Pregnancy
- Active malignancy
- Active immunosuppression therapy
- Kidney transplantation history
- Concomitant glomerulonephritis
- Nephrotic syndrome
- Solid organ transplantation history
- Autosomal dominant or recessive polycystic kidney disease
- Known renovascular disease
- Kidney failure (hemodialysis, peritoneal dialysis, or kidney transplantation)
- Active tobacco use
- Body weight >150 kg or BMI >50
- Uncontrolled hypertension: systolic blood pressure (SBP) >180 mmHg despite antihypertensive therapy
- Recent cardiovascular event (myocardial infarction, stroke, congestive heart failure within 6 months)
- Evidence of hepatitis B or C, or HIV infection, chronic
- Anticoagulation therapy requiring heparin bridging for procedures
- History of methicillin-resistant *Staphylococcus aureus* colonization
- Recent plastic, chemical, or surgical manipulation of adipose tissue for cosmetic purposes within 6 months
- Inability to provide informed consent
- Potentially unreliable subjects and those judged by the investigator to be unsuitable for the study

Diabetes Mellitus: Type 2 Diabetes (China)

Chapter 31
Efficacy and Safety of Umbilical Cord Mesenchymal Stem Cell Transplantation in Patients with Type 2 Diabetes Mellitus

Brief Summary:

This study is a 24-week single-center, randomized, double-blind, placebo-controlled trial. The trial includes a 3-week early screening and lifestyle education period, 6-week treatment period, and 18-week follow-up period. Chinese type 2 diabetic subjects receiving traditional hypoglycemic treatment were randomly assigned to umbilical cord mesenchymal stem cell or placebo infusion therapy to observe the efficacy and safety of umbilical cord mesenchymal stem cell infusion therapy.

Condition or disease	Intervention/treatment	Phase
Type 2 diabetes Mesenchymal stem cells	Biological: Umbilical cord mesenchymal stem cells Biological: Saline	Phase 1 Phase 2

Detailed Description:

This study is a 24-week single-center, randomized, double-blind, placebo-controlled trial. The trial includes a 3-week early screening and lifestyle education period, 6-week treatment period, and 18-week follow-up period. Chinese type 2 diabetic subjects receiving traditional hypoglycemic treatment (including insulin) were randomly assigned to umbilical cord mesenchymal stem cell or placebo infusion therapy to observe the efficacy and safety of umbilical cord mesenchymal stem cell infusion therapy.

Study Type:	Interventional (Clinical Trial)
Estimated Enrollment:	30 Participants
Allocation:	Randomized
Intervention Model:	Parallel Assignment

Shanghai East Hospital, Shanghai Tongji University, Shanghai, Shanghai, China
ClinicalTrials.gov Identifier: NCT04441658

Actual Study Start Date:	April 10, 2020
Estimated Primary Completion Date:	July 2022
Estimated Study Completion Date:	December 2022

Arm	Intervention/treatment
Experimental: Experimental group The volunteers of the experimental group will be given peripheral intravenously a dose of 0.75*10^6/kg of human umbilical cord mesenchymal stem cells at weeks 0, 1, 5, and 6	Biological: Umbilical cord mesenchymal stem cells Human umbilical cord mesenchymal stem cells were peripheral intravenously infused to the experimental group
Placebo comparator: control group The control group will be given the same dose of saline containing human albumin	Biological: Saline Saline containing human albumin will be infused to the control group

Primary Outcome Measures:

1. The changes in the HbA1C level [time frame: 24 weeks after treatment]

    The changes in the HbA1C level after transplantation

Secondary Outcome Measures:

1. Reduction of insulin requirement [time frame: 24 weeks after treatment]

    Reduction of insulin requirement after transplantation

2. The changes in blood glucose level [time frame: 24 weeks after treatment]

    The changes in blood glucose level after transplantation

Ages Eligible for Study:	30 Years to 75 Years (Adult, Older Adult)
Sexes Eligible for Study:	All
Accepts Healthy Volunteers:	No

Criteria

Inclusion Criteria:

1. Type 2 diabetes
2. The course of diabetes is 5–15 years
3. 20 kg/m^2 ≤ body mass index (BMI) ≤30 kg/m^2
4. 7.5% ≤ HbA1c ≤10%
5. Insulin dose and other oral antidiabetic drug doses should be stable within 3 months prior to randomization

Exclusion Criteria:

1. Heavy allergic constitution or an allergy to any component used in cell culture.
2. Being treated with drugs (glucocorticoids, tricyclic antidepressive agents, etc.) that affect the metabolism of glucose in the past 1 month.
3. Other causes of diabetes.
4. All kinds of acute complications such as diabetic ketoacidosis and non-ketohyperosmotic syndrome were screened in the past 6 months.
5. Having evidence of ongoing or frequent/severe hypoglycemia in the past 6 months.
6. Severe cardiovascular and cerebral events: occurrence of heart failure NYHA Classification III or IV, myocardial infarction, cerebral infarction, and cerebral hemorrhage within 6 months before the observation period.
7. Patients with abnormal liver and kidney function: AST and ALT exceeding 2.5 times the normal upper limit, and serum creatinine exceeding 1.5 mg/dL for men, exceeding 1.4 mg/dL for women.
8. Positive results of HbsAg, Anti-HCV, HIV, or syphilis.
9. Patients suffering from other serious systemic diseases (such as malignancy, central nervous system, cardiovascular system, blood system, digestive system, endocrine system, respiratory system, genitourinary system, immune system).
10. Ongoing pregnancy or absence of effective contraception in women with child-bearing potential.
11. Patients who had received other stem cell therapy before screening.

Chapter 32
Stem Cell Educator Therapy in Diabetes

Brief Summary:
Stem cell educator (SCE) therapy circulates a patient's blood through a blood cell separator, briefly co-cultures the patient's immune cells with adherent cord blood stem cells (CB-SCs) in vitro, and returns only the "educated" autologous immune cells to the patient's circulation. Several mechanistic studies with clinical samples and animal models have demonstrated the proof of concept and clinical safety of the SCE therapy. They suggest that the SCE therapy may function via CB-SC induction of immune tolerance in the autoimmune T cells and pathogenic monocytes/macrophages when these are exposed to the autoimmune regulator protein (AIRE) in the CB-SCs. In this project, the optimized SCE therapy for type 1 diabetes (T1D) and T2D will be tested in a prospective, single-arm, open-label, single-center study to assess its clinical efficacy and the related molecular mechanisms in patients with diabetes.

Condition or disease	Intervention/treatment	Phase
Diabetes mellitus, type 1 Diabetes mellitus, type 2	Combination product: Stem cell educator therapy	Phase 2

Study Type:	Interventional (Clinical Trial)
Estimated Enrollment:	100 Participants
Allocation:	N/A
Intervention Model:	Single- Group Assignment
Intervention Model Description:	In this project, the optimized SCE therapy for diabetes will be tested in a prospective, single-arm, open-label, single-center study

Department of Endocrinology, Chinese PLA General Hospital, Beijing, China
ClinicalTrials.gov Identifier: NCT03390231

© The Author(s), under exclusive license to Springer Nature
Switzerland AG 2022
J. N. Weiss, *Stem Cell Surgery Trials in Heart Failure and Diabetes*,
https://doi.org/10.1007/978-3-030-78010-4_32

Actual Study Start Date:	November 27, 2017
Estimated Primary Completion Date:	December 30, 2020
Estimated Study Completion Date:	December 31, 2020

Arm	Intervention/treatment
Experimental: Stem cell educator The stem cell educator (SCE) technology involves a closed-loop system that circulates a patient's blood through a blood cell separator, briefly co-cultures the patient's immune cells with adherent CB-SCs in vitro, and returns only the "educated" immune cells to the patient's circulation. Several mechanistic studies with clinical samples and animal models have been conducted to demonstrate the proof of concept and clinical safety of the SCE therapy. They suggest that the SCE therapy may function via CB-SC induction of immune tolerance in the autoimmune T cells and pathogenic monocytes/macrophages that are encountered through the action of the autoimmune regulator (AIRE) and other molecular mechanisms. Following induction of immune tolerance in the immune cells, the immune balance and homeostasis may be restored when treated cells are returned in vivo	Combination product: Stem cell educator therapy It briefly co-cultures the patient's lymphocytes with CB-SCs in vitro, induces immune tolerance through the action of autoimmune regulator (AIRE, expressed by CB-SCs), returns the educated autologous lymphocytes to the patient's circulation, and restores immune balance and homeostasis

Primary Outcome Measures:

Changes in inflammation-related markers in diabetic patients after stem cell educator therapy [time frame: 30 days]

After treatment for 30 days, diabetic patients will be tested for inflammation-related markers (e.g., Th1/Th2 cytokines) by flow cytometry and compared with the baseline levels.

Secondary Outcome Measures:

Change in insulin resistance [time frame: 30 days]

Before treatment, test for insulin sensitivity by chip analysis as baseline; after treatment, test sensitivity levels on the 1st month.

Metabolic control in the HbA1C levels [time frame: 3 months]

Before treatment, test for the HbA1C levels as baseline; after treatment, repeat these testing on the 3rd month. Hemoglobin A1c (HbA1c) will be reported as changes in percentage.

Metabolic control in the blood glucose levels [time frame: 3 months]

Before treatment, test for fasting glucose as baseline; after treatment, repeat these testing on the 3rd month. The blood glucose level will be measured as mMol/L.

Metabolic control in fasting C-peptide levels [time frame: 3 months]

Before treatment, test for C-peptide levels as baseline; after treatment, repeat these testing on the 3rd month. C-peptide will be measure as ng/ml.

Ages Eligible for Study:	20 Years to 60 Years (Adult)
Sexes Eligible for Study:	All
Accepts Healthy Volunteers:	No

Criteria

Inclusion Criteria:
- Patients are screened for enrollment in the study if both clinical signs and laboratory tests meet the diagnosis standards of the American Diabetes Association.

Exclusion Criteria:
- Exclusion criteria are any clinically significant diseases in the liver, kidney, and heart. Additional exclusion criteria are pregnancy, immunosuppressive medication, viral diseases, or diseases associated with immunodeficiency.

Publications of Results:

Delgado E, Perez-Basterrechea M, Suarez-Alvarez B, Zhou H, Revuelta EM, Garcia-Gala JM, Perez S, Alvarez-Viejo M, Menendez E, Lopez-Larrea C, Tang R, Zhu Z, Hu W, Moss T, Guindi E, Otero J, Zhao Y. Modulation of autoimmune T-cell memory by stem cell educator therapy: phase 1/2 clinical trial. EBioMedicine. 2015;2(12):2024–36.   https://doi.org/10.1016/j.ebiom.2015.11.003.   eCollection 2015.

Zhao Y, Jiang Z, Delgado E, Li H, Zhou H, Hu W, Perez-Basterrechea M, Janostakova A, Tan Q, Wang J, Mao M, Yin Z, Zhang Y, Li Y, Li Q, Zhou J, Li Y, Martinez Revuelta E, Maria García-Gala J, Wang H, Perez-Lopez S, Alvarez-Viejo M, Menendez E, Moss T, Guindi E, Otero J. Platelet-derived mitochondria display embryonic stem cell markers and improve pancreatic islet β-cell function in humans. Stem Cells Transl Med. 2017;6(8):1684–97. https://doi.org/10.1002/sctm.17-0078. Epub 2017 Jul 7.

Zhao Y, Jiang Z, Zhao T, Ye M, Hu C, Yin Z, Li H, Zhang Y, Diao Y, Li Y, Chen Y, Sun X, Fisk MB, Skidgel R, Holterman M, Prabhakar B, Mazzone T. Reversal of type 1 diabetes via islet β cell regeneration following immune modulation by cord blood-derived multipotent stem cells. BMC Med. 2012;10:3. https://doi.org/10.1186/1741-7015-10-3

Zhao Y, Jiang Z, Zhao T, Ye M, Hu C, Zhou H, Yin Z, Chen Y, Zhang Y, Wang S, Shen J, Thaker H, Jain S, Li Y, Diao Y, Chen Y, Sun X, Fisk MB, Li H. Targeting insulin resistance in type 2 diabetes via immune modulation of cord blood-derived multipotent stem cells (CB-SCs) in stem cell educator therapy: phase I/II clinical trial. BMC Med. 2013;11:160. https://doi.org/10.1186/1741-7015-11-160

Chapter 33
Clinical Study of Umbilical Cord Mesenchymal Stem Cells in the Treatment of Type 2 Diabetic Nephropathy

Brief Summary:

The study was a 96-week multicenter, randomized, double-blind, placebo-controlled trial that included 3 weeks of pre-screening and lifestyle education, 32 weeks of treatment, and 64 weeks of follow-up. This study was conducted to observe the efficacy and safety of umbilical cord mesenchymal stem cell infusion in Chinese patients with type 2 diabetic nephropathy who received traditional hypoglycemic therapy.

Condition or disease	Intervention/treatment	Phase
Type 2 diabetes with renal manifestations	Biological: Human umbilical cord mesenchymal stem cells Other: Saline	Phase 2

Detailed Description:

The study was a 96-week multicenter, randomized, double-blind, placebo-controlled trial that included 3 weeks of pre-screening and lifestyle education, 32 weeks of treatment, and 64 weeks of follow-up. We plan to recruit 54 subjects, which will be divided into the experimental group and control group. The volunteers of the experimental group will be given peripheral intravenously a dose of $1.5 * 10^6$/kg of human umbilical cord mesenchymal stem cell (HUC-MSC) at weeks 0, 8, 16, 24, and 32. The control group will be given the same dose of saline. Then centralization visit will be conducted every 12 weeks until the 96th week. The primary endpoints include estimated glomerular filtration rate and urinary albumin creatinine ratio (UACR). The secondary endpoints include HbA1C, plasma insulin and C-peptide, and insulin dose.

Shanghai East Hospital, Shanghai, Shanghai, China
ClinicalTrials.gov Identifier: NCT04216849

© The Author(s), under exclusive license to Springer Nature Switzerland AG 2022

J. N. Weiss, *Stem Cell Surgery Trials in Heart Failure and Diabetes*, https://doi.org/10.1007/978-3-030-78010-4_33

Study Type:	Interventional (Clinical Trial)
Estimated Enrollment:	54 Participants
Allocation:	Randomized
Intervention Model:	Parallel Assignment

Actual Study Start Date:	November 1, 2019
Estimated Primary Completion Date:	November 1, 2021
Estimated Study Completion Date:	November 1, 2022

Arm	Intervention/treatment
Experimental: experimental group The volunteers of the experimental group will be given peripheral intravenously a dose of $1.5 * 10^6$/kg of human umbilical cord mesenchymal stem cells at weeks 0, 8, 16, 24, and 32	Biological: Human umbilical cord mesenchymal stem cells Human umbilical cord mesenchymal stem cells were peripheral intravenously infused to the experimental group
Placebo comparator: Control group The control group will be given the same dose of saline containing human albumin	Other: Saline Saline containing human albumin will be infused to the control group

Primary Outcome Measures:

1. eGFR [time frame: 96 weeks after treatment]

   Estimated glomerular filtration rate

2. UACR [time frame: 96 weeks after treatment]

   Urinary albumin creatinine ratio

Ages Eligible for Study:	30 Years to 70 Years (Adult, Older Adult)
Sexes Eligible for Study:	All
Accepts Healthy Volunteers:	No

Criteria

Inclusion Criteria:
- Type 2 diabetes mellitus
- $7.5\% \leq$ HbA1C $\leq 10\%$
- UACR ≥ 30 mg/g creatinine
- 30 mL/min/1.73 m^2 $\leq$ eGFR <60/milliliter/1.73 m^2

Exclusion Criteria:
- Type 1 diabetes mellitus
- Tumor history
- Other causes of chronic kidney disease
- Abnormal liver function

Part XIV
Diabetes Mellitus: Type 2 Diabetes (Near East)

Chapter 34
BM-MNC and UCMSC for Type 2 Diabetes Mellitus Patients

Brief Summary:
The aim of this preliminary study is to evaluate the safety and efficacy of bone marrow mononuclear cells (BM-MNCs) and umbilical cord tissue-derived mesenchymal stem cell (UC-MSCs) administration in type 2 diabetes patients.

Condition or disease	Intervention/treatment	Phase
T2D	Biological: Bone-marrow aspiration, intra-pancreatic catheterization of BM-MNC Biological: Intravenous infusion of UC-MSC	Phase 1 Phase 2

Detailed Description:
Type 2 diabetes (T2D) patients had peripheral insulin resistance accompanied by progressive pancreatic beta-cell degeneration and dysfunction due to glucotoxicity and lipotoxicity. Several studies have shown that the immune system plays a significant role in the pathogenesis of T2D. Bone marrow mononuclear cells (BM-MNCs) and umbilical cord tissue-derived mesenchymal stem cells (UC-MSCs) via its immunomodulatory properties have the potential to improve insulin resistance condition and pancreatic beta-cell dysfunction, thus improving the glycemic control and insulin requirement in T2D patients. In this pilot study, we plan to recruit 15 T2D patients with total daily dose of insulin ≥ 0.5 unit/kgBW/day to receive BM-MNC (5 subjects) or UC-MSC injections (10 subjects). These subjects will be closely followed up for 12 months for the evaluation of primary and secondary outcomes.

Faculty of Medicine, Universitas Indonesia, Jakarta Pusat, DKI Jakarta, Indonesia
ClinicalTrials.gov Identifier: NCT04501341

© The Author(s), under exclusive license to Springer Nature Switzerland AG 2022
J. N. Weiss, *Stem Cell Surgery Trials in Heart Failure and Diabetes*,
https://doi.org/10.1007/978-3-030-78010-4_34

Study Type:	Interventional (Clinical Trial)
Estimated Enrollment:	15 Participants
Allocation:	Non-randomized
Intervention Model:	Parallel Assignment

Actual Study Start Date:	March 14, 2016
Estimated Primary Completion Date:	December 1, 2021
Estimated Study Completion Date:	December 1, 2021

Arm	Intervention/treatment
Experimental: BM-MNC experimental Autologous bone marrow mononuclear cell	Biological: Bone marrow aspiration, intra-pancreatic catheterization of BM-MNC Autologous bone marrow mononuclear cells infused to the main blood vessels that supply the pancreas according to the results of previous pancreatic CT-scan, performed by an interventional radiologist. The target is to distribute the BM-MNCs equally to all parts of the pancreas. Dosage: 1×10^5–1×10^6 CD34 cells/kgBW
Experimental: UC-MSC Umbilical cord mesenchymal stem cell	Biological: Intravenous infusion of UC-MSC Allogeneic umbilical cord tissue-derived mesenchymal stem cells will be given via intravenous infusion. Dosage: 2×10^6 cells/kgBW, twice, with a 3-month interval

Primary Outcome Measures:
1. Decreasing total daily dose of insulin ($\geq 30\%$) [time frame: before intervention and the 1st, 3rd, 6th, and 12th month after intervention]

After intervention, blood glucose level will be reported by the subjects on a weekly basis. The insulin dose and/or oral medication will be adjusted accordingly.

Secondary Outcome Measures:
1. Increasing C-peptide level [time frame: before intervention and the 1st, 3rd, 6th, and 12th month after intervention]

Measurements were obtained with mixed meal tolerance test

2. Decreasing insulin resistance level [time frame: before intervention and the 1st, 3rd, 6th, and 12th month after intervention]

Measurement of HOMA-IR, calculated using fasting C-peptide and fasting plasma glucose formula

3. Immunology/inflammatory markers [time frame: before intervention and the 1st, 3rd, 6th, and 12th month after intervention]

Measurements of Interleukin-10 and TNF-alpha from the serum and supernatant from PBMC stimulation

4. Adverse events [time frame: up to 12 months after intervention]

Thrombosis, hemorrhage, and infection

5. HbA1c [time frame: before intervention and the 1st, 3rd, 6th, and 12th month after intervention]

Stable HbA1c or decreasing HbA1c (from baseline)

Ages Eligible for Study:	30 Years to 65 Years (Adult, Older Adult)
Sexes Eligible for Study:	All
Accepts Healthy Volunteers:	No

Criteria

Inclusion Criteria:
- Type 2 diabetes patients on insulin therapy with or without oral hypoglycemic agents, with total daily dose of insulin ≥0,5 unit/kg body weight
- Stable HbA1C in the last 6six months (HbA1c ≤8.5%)

Exclusion Criteria:
- Type 1 diabetes mellitus
- eGFR <45 mL/min/m^2 (for BM-MNC)
- Liver disease (moderate–severe)
- Active infection
- Contrast hypersensitivity (for BM-MNC)
- History of malignancy
- Acute coronary syndrome in the last 3 months
- Coronary arterial diseases with significant stenosis and has not carried out revascularization
- Pregnancy (for women subjects)

Publications of Results:
Gao F, Chiu SM, Motan DA, Zhang Z, Chen L, Ji HL, Tse HF, Fu QL, Lian Q. Mesenchymal stem cells and immunomodulation: current status and future prospects. Cell Death Dis. 2016;7:e2062. https://doi.org/10.1038/cddis.2015.327. Review.

Weiss ARR, Dahlke MH. Immunomodulation by mesenchymal stem cells (MSCs): mechanisms of action of living, apoptotic, and dead MSCs. Front Immunol. 2019;10:1191. https://doi.org/10.3389/fimmu.2019.01191. eCollection 2019. Review.

Other Publications:

Bhansali A, Asokumar P, Walia R, Bhansali S, Gupta V, Jain A, Sachdeva N, Sharma RR, Marwaha N, Khandelwal N. Efficacy and safety of autologous bone marrow-derived stem cell transplantation in patients with type 2 diabetes mellitus: a randomized placebo-controlled study. Cell Transplant. 2014;23(9):1075–85.

Campbell RK, Martin TM. The chronic burden of diabetes. Am J Manag Care. 2009;15(9 Suppl):S248–54.

Chao YH, Wu HP, Chan CK, Tsai C, Peng CT, Wu KH. Umbilical cord-derived mesenchymal stem cells for hematopoietic stem cell transplantation. J Biomed Biotechnol. 2012;2012:759503. https://doi.org/10.1155/2012/759503. Epub 2012 Oct 3. Review.

Estrada EJ, Valacchi F, Nicora E, Brieva S, Esteve C, Echevarria L, Froud T, Bernetti K, Cayetano SM, Velazquez O, Alejandro R, Ricordi C. Combined treatment of intrapancreatic autologous bone marrow stem cells and hyperbaric oxygen in type 2 diabetes mellitus. Cell Transplant. 2008;17(12):1295–304.

Féry F, Paquot N. [Etiopathogenesis and pathophysiology of type 2 diabetes]. Rev Med Liege. 2005;60(5–6):361–8. Review. French.

Guan LX, Guan H, Li HB, Ren CA, Liu L, Chu JJ, Dai LJ. Therapeutic efficacy of umbilical cord-derived mesenchymal stem cells in patients with type 2 diabetes. Exp Ther Med. 2015;9(5):1623–30. Epub 2015 Mar 9.

Hu J, Li C, Wang L, Zhang X, Zhang M, Gao H, Yu X, Wang F, Zhao W, Yan S, Wang Y. Long term effects of the implantation of autologous bone marrow mononuclear cells for type 2 diabetes mellitus. Endocr J. 2012;59(11):1031–9. Epub 2012 Jul 13.

Itariu BK, Stulnig TM. Autoimmune aspects of type 2 diabetes mellitus – a minireview. Gerontology. 2014;60(3):189-96. https://doi.org/10.1159/000356747. Epub 2014 Jan 22. Review.

Tsai S, Clemente-Casares X, Revelo XS, Winer S, Winer DA. Are obesity-related insulin resistance and type 2 diabetes autoimmune diseases? Diabetes. 2015;64(6):1886–97. https://doi.org/10.2337/db14-1488. Review.

Wehbe T, Chahine NA, Sissi S, Abou-Joaude I, Chalhoub L. Bone marrow derived stem cell therapy for type 2 diabetes mellitus. Stem Cell Invest. 2016;3:87. https://doi.org/10.21037/sci.2016.11.14. eCollection 2016.

Part XV
Diabetes Mellitus: Type 2 Diabetes (Vietnam)

Chapter 35
BM-MNC and UC-MSC Infusion for Type 2 Diabetes Mellitus Patients (T2DM)

Brief Summary:

The purpose of this study is to evaluate the preliminary safety and efficacy of autologous bone marrow mononuclear cell (BM-MNCs) and allogeneic umbilical cord tissue-derived mesenchymal stem cell (UC-MSC) infusion in type 2 diabetes mellitus patients.

Condition or disease	Intervention/treatment	Phase
Type 2 Diabetes mellitus	Biological: BM-MNC and UC-MSC Other: Control	Phase 1 Phase 2

Detailed Description:

Mononuclear cells are collected from autologous bone marrow, and allogeneic mesenchymal stem cells are isolated and cultured from umbilical cord tissues.

Thirty patients with type 2 diabetes mellitus will be enrolled and will receive mononuclear cell and mesenchymal stem cell by intravenous infusion and followed up for 6 months. The other 30 patients with type 2 diabetes mellitus will be enrolled and treated with standard medicines, which would be used as the control group.

Safety is to assess the occurrence of adverse events (AEs) during either stem cell infusion or by physician assessments. The primary endpoint is to assess the improvement of patient's C-peptide and HOMA-β, HOMA-IR, cytokines TNF-α and IL-1β, blood glucose level, and hemoglobin A1c (HbA1c) level.

Van Hanh General Hospital, Ho Chi Minh City, Ho Chi Minh, Vietnam
ClinicalTrials.gov Identifier: NCT03943940

© The Author(s), under exclusive license to Springer Nature Switzerland AG 2022
J. N. Weiss, *Stem Cell Surgery Trials in Heart Failure and Diabetes*,
https://doi.org/10.1007/978-3-030-78010-4_35

Study Type:	Interventional (Clinical Trial)
Estimated Enrollment:	60 Participants
Allocation:	Non-randomized
Intervention Model:	Parallel Assignment

Actual Study Start Date:	April 24, 2019
Estimated Primary Completion Date:	December 30, 2019
Estimated Study Completion Date:	August 30, 2020

Arm	Intervention/treatment
Experimental: BM-MNC and UC-MSC 30 patients with type 2 diabetes mellitus will be enrolled and will receive mononuclear cells and mesenchymal stem cells by intravenous infusion	Biological: BM-MNC and UC-MSC Autologous bone marrow mononuclear cells (BM-MNCs) and allogeneic umbilical cord tissue-derived mesenchymal stem cells (UC-MSCs) under sterile conditions to treat this disease UC-MSC: $1–2 \times 10^6$ cells/kg
Standard medicines 30 patients with type 2 diabetes mellitus will be enrolled and treated with standard medicines	Other: Control Standard medicine

Primary Outcome Measures:

1. The levels of C-peptide and HOMA-β [time frame: enrollment and 1 month, 3 months, and 6 months after transplantation]

   Assess the changes in C-peptide and HOMA-β levels after transplantation

2. The levels of HOMA-IR and cytokines TNF-α and IL-1β [time frame: enrollment and 1 month, 3 months, and 6 months after transplantation]

   Assess the changes in the levels of HOMA-IR and cytokines TNF-α and IL-1β after transplantation

3. Blood glucose level [time frame: enrollment and 1 month, 3 months, and 6 months after transplantation]

   Assess the changes in the blood glucose level after transplantation

4. Hemoglobin A1c (HbA1c) level [time frame: enrollment and 1 month, 3 months, and 6 months after transplantation]

   Assess the changes in the HbA1C level after transplantation

5. Adverse events [time frame: during the course of 6 months]

   Number of adverse events in both groups

Secondary Outcome Measures:
1. Insulin dose and drug dosage [time frame: enrollment and 1 month, 3 months, and 6 months after transplantation]

   Assess the changes in insulin dose and drug dosage after transplantation

Ages Eligible for Study:	18 Years to 70 Years (Adult, Older Adult)
Sexes Eligible for Study:	All
Accepts Healthy Volunteers:	No

Criteria

Inclusion Criteria:
- Who is diagnosed with type 2 diabetes mellitus according to the ADA for 3 years or more
- Able to read, write, and understand the ICF form and agree to participate in the study
- Males and females between 18 and 70 years old at the time of screening
- FBG >7 mmol/L
- $8\% \leq HbA1C \leq 11\%$
- Fasting C-peptide >0.6 ng/mL
- Anti-GAD (−)
- Treated with two oral diabetes medications but has uncontrolled blood glucose (HbA1C $\geq$8%)

Exclusion Criteria:
- Pregnant women, women planning to become pregnant, and lactating women during the study period
- The patient has a disease or a history of vascular disease; history of abdominal or chest aortic disease
- Patients diagnosed with heart failure class IV according to the NYHA or kidney failure class IV according to the KDIGO
- Patients with severe malignancy or dysplasia within 5 years prior to the study period or who are suffering from severe malignant or dysplasia
- Has infection and undergoing antibiotic treatment or has discontinued antibiotic treatment within 14 days
- Hematologic disease or coagulopathy
- With abnormalities in the liver function (AST and/or ALT $\geq$2 times or bilirubin $\geq$2.0 times the normal value at the time of screening)
- With immunodeficiency diseases such as HIV or hepatitis B and C
- Acute or chronic pancreatitis or a history of acute pancreatitis
- Patients taking immunosuppressive drugs (such as azathioprine, methotrexate) within 6 months before the study time or taking immunosuppressive drugs
- Unable to complete the study
- Participation in another study

Part XVI
Other Thoughts

Chapter 36
Conclusion

Diabetes mellitus has traditionally been classified as either type 1 or type 2, on the basis of the age of diagnosis and the presence of autoantibodies to pancreatic islet β-cell antigens. A third type of diabetes has been identified, namely, "latent autoimmune diabetes in adults," which is diagnosed by the presence of glutamic acid decarboxylase antibodies. Difficulties in diabetes classification leads to problems in outcome determinations. The possibility of genetic classification may result in more accurate diagnosis and prognosis.

Type 1 Diabetes

Since the autoimmune destruction of islet β-cell antigens leads to type 1 diabetes, the preservation of these cells is important. Islet β-cells have limited regenerative potential; therefore attention has been directed to the reprogramming of other cells, specifically α-cells, also located within the pancreatic islet and similarly developed.

Work in rodents [1, 2] demonstrated that spontaneous insulin production by α-cells is dynamically regulated to cope with insulin insufficiency. Both α-cells and γ cells were found to be capable of reprogramming to produce and secrete insulin in response to the changing glucose. The modified α-cells were less likely to be immune response-affected than β-cells. This work demonstrates the plasticity of islet cells.

Dirice et al. [3] demonstrated that increasing the proliferation of β-cells prior to immunologic attack provides protection against the development of type 1 diabetes.

Type 2 Diabetes

Ahlqvist et al. [4] performed a cluster analysis of newly diagnosed diabetes using glutamate decarboxylase antibodies, age at diagnosis, body mass index (BMI), glycated hemoglobin (HgbA1c), and homeostatic model assessment estimates of β-cell

© The Author(s), under exclusive license to Springer Nature
Switzerland AG 2022
J. N. Weiss, *Stem Cell Surgery Trials in Heart Failure and Diabetes*,
https://doi.org/10.1007/978-3-030-78010-4_36

function and insulin resistance as well as prospective patient record complications and prescription data. Five novel subgroups of adult-onset diabetes in relation to outcomes were identified:

1. Severe autoimmune diabetes (SAID)
2. Severe insulin-deficient diabetes (SIDD)
3. Severe insulin-resistant diabetes (SIRD)
4. Mild obesity-related diabetes (MOD)
5. Mild age-related diabetes (MARD)

Utilization of this metric may lead to precision medicine in diabetes treatments.

Obesity is a major risk factor for the development of type 2 diabetes. Lotta et al. [5] found that variants in the MC4R gene were associated with a decrease in type 2 diabetes, coronary artery disease, lower BMI, and obesity. This work has implications for the protection against developing obesity and its attendant complications.

Studies of insulin signaling by Hancock et al. [6] demonstrated a non-canonical pathway regulating target genes impacting insulin-related functions. Insulin receptor translocation is a novel pathway that may lead to further discoveries.

Riddle et al. [7] work's on *A. mexicanus*, a cave-dwelling fish, found that despite having high fasting blood glucose and insulin resistance, from a mutation in the insulin receptor, the fish remained healthy and exhibited a normal life span. The authors postulate that there are compensatory mechanisms in a nutrient-poor environment. This work may lead to the development of treatments for hyperglycemia-related diabetic complications.

Diet affects the gut microbiome, which may modify the sensitivity to insulin. Koh et al. [8] discovered that imidazole propionate, a gut microbiota-dependent metabolite, is related to the development of insulin resistance. This knowledge may lead to the development of pharmacologic targets for treatment.

References

1. Cigliola V, Ghila L, Thorel F, van Gurp L, Baronnier D, Oropeza D, Gupta S, Miyatsuka T, Kaneto H, Magnuson MA, Osipovich AB, Sander M, Wright CEV, Thomas MK, Furuyama K, Chera S, Herrera PL. Pancreatic islet-autonomous insulin and smoothened-mediated signalling modulate identity changes of glucagon(+) alpha-cells. Nat Cell Biol. 2018;20:1267–77. https://doi.org/10.1038/s41556-018-0216-y.
2. Furuyama K, Chera S, van Gurp L, Oropeza D, Ghila L, Damond N, Vethe H, Paulo JA, Joosten AM, Berney T, Bosco D, Dorrell C, Grompe M, Ræder H, Roep BO, Thorel F, Herrera PL. Diabetes relief in mice by glucose-sensing insulin-secreting human alpha-cells. Nature. 2019;567:43–8. https://doi.org/10.1038/s41586-019-0942-8.
3. Dirice E, Kahraman S, De Jesus DF, et al. Increased β-cell proliferation before immune cell invasion prevents progression of type 1 diabetes. Nat Metab. 2019;1:509–18. https://doi.org/10.1038/s42255-019-0061-8.

4. Ahlqvist E, Storm P, Karajamaki A, et al. Novel subgroups of adult-onset diabetes and their association with outcomes: a data-driven cluster analysis of six variables. Lancet Diabetes Endocrinol. 2018;6:361–9. https://doi.org/10.1016/S2213-8587(18)30051-2.
5. Lotta LA, Mokrosinski J, Mendes de Oliveira E, et al. Human gain-of-function mc4r variants show signaling bias and protect against obesity. Cell. 2019;177:597–607. e599. https://doi.org/10.1016/j.cell.2019.03.044.
6. Hancock ML, Meyer RC, Mistry M, Khetani RS, Wagschal A, Shin T, Ho Sui SJ, Näär AM, Flanagan JG. Insulin receptor associates with promoters genome-wide and regulates gene expression. Cell. 2019;177:722–36. e722. https://doi.org/10.1016/j.cell.2019.02.030.
7. Riddle MR, Aspiras AC, Gaudenz K, Peuß R, Sung JY, Martineau B, Peavey M, Box AC, Tabin JA, McGaugh S, Borowsky R, Tabin CJ, Rohner N. Insulin resistance in cavefish as an adaptation to a nutrient-limited environment. Nature. 2018;555:647–51. https://doi.org/10.1038/nature26136.
8. Koh A, Molinaro A, Stahlman M, et al. Microbially produced imidazole propionate impairs insulin signaling through mtorc1. Cell. 2018;175:947–61. e917. https://doi.org/10.1016/j.cell.2018.09.055.

Index

Printed in the United States
by Baker & Taylor Publisher Services

Printed in the United States
by Baker & Taylor Publisher Services